Indian Statistical Institute Series

The **Indian Statistical Institute Series,** a Scopus-indexed series, publishes high-quality content in the domain of mathematical sciences, bio-mathematics, financial mathematics, pure and applied mathematics, operations research, applied statistics and computer science and applications with primary focus on mathematics and statistics. Editorial board comprises of active researchers from major centres of the Indian Statistical Institute.

Launched at the 125th birth Anniversary of P.C. Mahalanobis, the series will publish high-quality content in the form of textbooks, monographs, lecture notes, and contributed volumes. Literature in this series are peer-reviewed by global experts in their respective fields, and will appeal to a wide audience of students, researchers, educators, and professionals across mathematics, statistics and computer science disciplines.

More information about this series at https://link.springer.com/bookseries/15910

Arijit Chaudhuri · Sanghamitra Pal

A Comprehensive Textbook on Sample Surveys

 Springer

Arijit Chaudhuri
Applied Statistics Unit
Indian Statistical Institute
Kolkata, India

Sanghamitra Pal
Department of Statistics
West Bengal State University
Kolkata, India

ISSN 2523-3114   ISSN 2523-3122 (electronic)
Indian Statistical Institute Series
ISBN 978-981-19-1420-1   ISBN 978-981-19-1418-8 (eBook)
https://doi.org/10.1007/978-981-19-1418-8

Cover photo: Reprography & Photography Unit, Indian Statistical Institute, Kolkata, India

This Springer imprint is published by the registered company Springer Nature Singapore Pte Ltd.
The registered company address is: 152 Beach Road, #21-01/04 Gateway East, Singapore 189721,
Singapore

To
Bulu
from Arijit Chaudhuri

To my Late Parents
Sanghamitra Pal

Preface

The intended readers of this book are under-graduate, graduate and post-graduate students, their respective teachers, researchers in survey sampling and the practitioners at the government, corporate and private organizations. The authors feel it desirable to get a lucidly written comprehensive textbook on Sample Surveys.

The book starts at the elementary level comprehensible to the just school leavers with only background in simple arithmetic, algebra, geometry, rudimentary-level calculus and initiation in theory of probability at the classical level of calculations by using algebraic theories of permutation, combination and solution of the difference equations.

Concepts of population, sample, methods of sample selection, parameters and their estimation, concepts of statistics and their properties, standard errors of estimators and concepts of good samples, good estimators and number of samples to be drawn from a population with specific objectives to realize will all be introduced, explained and treated with appropriate motivations.

Initially, ramifications will be dealt with keeping an eye to the requirements of covering courses in sample surveys in many universities. Gradually, advanced motivations with theoretical niceties will progressively be laid before the readers. How research at desirably high level went on advancing in diverse directions will of course be recounted in course of the elucidation of the contents. But throughout, a keenness will be exposed to indicate how the theoretical motivations are dictated by requirements of practical applications in various ways.

Starting from the inference in survey sampling, the design-based and the model-assisted sampling theories in finite population setup are narrated elaborately in the book. Small-area estimation and the surveys of population containing rare units are also described according to the requirement of the post-graduate students and the research scholars. Sample-size determination is a vital issue in survey sampling. The determination of sample-sizes in complex strategies is also discussed in the book. In the Introduction, a detailed chapter-wise brief description is presented to acquaint our readers with what we really intend to equip them.

In order to estimate a finite population mean by a sample mean from a Simple Random Sample taken without Replacement (SRSWOR), a rational approach may be needed to demand the resulting error on either side bounded by a positive proper fraction f of the unknown population mean. The probability of that should be a high prescribed positive number as $1 - \alpha$ (α is positive but a small number, say 0.05). To achieve this, one may apply Chebyshev's inequality to prescribe the sample-size n to be tabulated in terms of the population-size N, f, α and specified values of population coefficients of variation. This approach naturally rationalizes choice of stratum-wise sample-sizes as well. Hence emerges a rational alternative to Neyman's (1934) allocation of sample-sizes in stratified SRSWOR—a simple point raised in our text. It is known that a Necessary and Sufficient Condition (NSC) for the existence of an unbiased estimator for a finite population total is the positivity of the inclusion-probability of every unit in the finite population. As a follow-up positivity of inclusion-probability of every distinct pair of finite population units is an NSC for the existence of an unbiased estimator of the variance of an unbiased estimator of a finite population total. Given sample data drawn from a complicated sampling design, how the gain in estimating a population mean or total from a comparable simpler design maybe estimated is also one of such simple survey sampling topics as are discussed in this book. Over the last four-five decades, small area estimation (SAE) is a hot topic of theoretical niceties and of enormous practical activities. Though we are not conspicuously active in this area since 1997, we cover here the essentials about this topic with an emphasis that data is covered by unequal probability sampling. A synthetic generalized regression estimation procedure substantially outperforms an initial Horvitz-Thompson estimator or a Rao-Hartley-Cochran estimator. A more sophisticated Empirical Bayes or Kalman Filter approach may not yield further efficacious results. Though Chaudhuri (2012) has dwelt on this, the modern researchers are not seen to shed more light on this.

To estimate household expenses, for example, on hospital treatment of inmates in the last one year for cancer, a usual household survey may lack enough information content, because many households may not contain such inmates at all, and there is a better alternative to take a sample of hospitals and of households, having inmates treated there of cancer last year. So, hospitals, as selection units, and households linked to them as observational units are to be sampled by a new sampling procedure, called *Network Sampling*, which is needed to be thoroughly studied as has been done in this text as an important practical requisite.

One shortcoming, however, in our present text is the absence of discussions on non-probability sampling procedures, like quota sampling, snowball sampling, web sampling procedures, etc. Some of them are dealt with in the text by Wu and Mary Thompson (2020). The reasons behind these are our lack of conviction of their efficacies and our effective participation in the concerned field.

A few more topics of popular interest, however, are also covered in our text, which we may not name or mention here as they are quite commonly known.

Arijit Chaudhuri gratefully acknowledges the help and encouragement received from the Director, Indian Statistical Institute (ISI) and his colleagues in the Applied Statistics Unit of ISI.

Sanghamitra Pal gratefully acknowledges the support and encouragement profusely received from her family.

Both the authors gratefully acknowledge laboriously prepared suggestions from the reviewers that helped improve the content.

Kolkata, India   Arijit Chaudhuri
                                                                        Sanghamitra Pal

References

Chaudhuri, A. (2012). *Developing small domain statistics – modeling in survey sampling* (e-book), Saarbrucken, Germany: Lambert Academic Publishing.
Wu, C. & Thompson, M. E. (2020). *Sampling theory and practice*, Springer.

About This Book

Our book is mainly intended to serve the needs for the undergraduate, graduate and post-graduate students of statistics, specifically survey sampling, their teachers in Indian universities, the private, government and corporate-level researchers handling statistical data to extract meaningful conclusions therefrom in judicious scientific manners. As by products of course the material presented may also serve corresponding needs for those outside India as well. Having profusely published in other text books, monographs and peer-reviewed journals in India and abroad on Survey Sampling over the last five decades it is our fond expectation to be able to cater to the intended needs through this venture in a desirable way.

Here we first modestly endeavour to explain succinctly the concepts of a finite survey population, statistics to be derived therefrom in suitable ways to bring out point and interval estimation procedures to produce inference procedures about parameters of interest to be appropriately defined to serve our purposeful interest.

We consider first classical, design-based inferential procedures, examine how they serve specific inferential purposes, how they are inadequate in their mission, then examine alternative approaches introducing model formulations, adopting predictive, model-design-based and model-assisted procedures of analysis, inference and interpretation. Next we consider Bayesian and Empirical Bayesian approaches, Complex procedures of stratification, clustering, sampling in multi stages and phases, resorting to linear and non-linear estimation of linear and non-linear parameters are dealt with. Network and adaptive sampling methods are covered. How to control sampling when inconvenient population units are encountered is also described. How unit-wise and item-wise non-response is to be dealt with is also covered. Problems for solution to test the readers' extent of assimilation are also presented with solutions or without solutions but with hints or without them. Some case studies with worked out details are presented too.

Introduction

The latest textbook on Survey Sampling, a major portion of which I (Arijit Chaudhuri) have recently gone through, is *Sampling Theory and Applications* by Prof. Raghunath Arnab published by Academic Press in 2017. It was really gigantic containing 26 chapters spread over more than 900 pages. This book, in fact, is an inducement for the present venture especially because the author is known to me for many years since his student days in the University of Calcutta and early academic career-building at the Indian Statistical Institute Kolkata, India. Though he is living now in South Africa for more than three decades, he of course has not left much gaps for us to fill in. But our intention is to strive to express ourselves rather in a concise manner to cater to the needs of those interested in Survey Sampling among the students, teachers, researchers and practitioners in India putting forward our personal motivations for the development of the subject.

More books, of course, have to be alluded to like *Theory and Methods of Survey Sampling* by Parimal Mukhopadhyay published by PHI Learning in 1998. Also, several important ones are authored by among others: Särndal, Swensson and Wretman (1992), Bolfarine and Zacks (1992), Valliant, Dorfman and Royall (2000), J. N. K. Rao (2003), J. N. K Rao and Molina (2015), Thompson S. K. (1992), Thompson M. E. (1997), Thompson and Seber (1996), Salehi and Seber (2013), Chaudhuri, Christofides and Rao (2016) besides those of which I am an author or a co-author. Almost, on concluding our present book, we happened to have a look at the book entitled *Sampling Theory and Practice* by Wu and Mary Thompson (2020). Though seemingly they somewhat mutually overlap, we stick to our project for reasons clearly announced at the outset.

Let me refer briefly to the consecutive chapters where we intend to narrate various topics we are inclined to cover. Chapter 1 gives a short overview of our conception of what must necessarily be grasped rather easily by our readers about to pry into the ins and outs of what messages the subject 'survey sampling' should convey. Before the concept of a 'sample' comes that of a 'population', the totality about which we are curious to know but may not be able to afford to do so. So, we are forced to be content with a glimpse of a part thereof which is a sample. We have to formally define both. The next question is how to get a sample and with what requirements. How and what story a sample tells us about the features of a population? How good, adequate or mis-representative is a sample at hand *vis-a-vis* the population it is to talk about? This scrutiny calls for defining parameters for populations, statistics and estimators to be defined for the samples. In this context, it is important to talk about 'randomness' and 'error' and how to 'control' the latter. These and related further simplistic issues are purported to be alluded to in our opening chapter. Certain exercises and lessons of relevance to test the reader's appreciation of the ideas hitherto exposed are given in the Lessons and Exercises section in Chap. 1.

In Chap. 2, essentially our purpose is to acquaint the reader with various approaches to the ways to proceed to utilize the samples towards assessing the features of the population, whence a sample has been judiciously drawn. Of course, it will be pointed out that various approaches are needed instead of trying to have a single way out. Quite a number of concepts for this purpose are needed to be discussed in clarifying why various alternative approaches are needed.

Chapter 3 elaborates requirements and procedures for varying probability sampling techniques. Stratification of a population, formation of clusters and selection in two or more stages and phases are explained accounting for the facilities they bring out in assessing the population features by dint of sampling in diverse ways. Why and how samples may be chosen on successive occasions in various ways to serve our purposes of inferring about populations are also spelled out with due emphasis.

In Chap. 4, we report how to fix the sample-size in equal probability sampling by using Chebyshev's inequality avoiding model postulation. Next, we discuss controlled sampling.

In Chap. 5, we discuss how to tackle unit non-response by weighting adjustments and item-non-response by the imputation technique.

In Chap. 6, we consider surveying for stigmatizing issues by indirect survey methods like Randomized Response (RR) Techniques (RRT) and other alternative adjustments.

In Chap. 7, we consider super-population modelling together with design-based methods for inference in finite population sampling. Also, we consider asymptotics in survey populations so as to employ model-assisted survey methods.

In Chap. 8, we consider prediction approach with modelling and robustness issues in case of model failures.

In Chap. 9, we treat Small Area Estimation (SAE) to improve upon poor estimators, because of inadequate sample-size by utilizing judicious modelling so as to borrow strength from like-domains without actual enhancement in sample-size. Samples are drawn by unequal probability sampling scheme. The Kalman filtering is also resorted to in order to borrow strength not only from related neighbouring areas but from past data bearing relationship of observations spread over time.

In Chap. 10, we consider treatment of non-linear parameters and non-linear functions of statistics by resorting to the Taylor series expansion method, Jack-knifing, Bootstrap and Balanced Repeated Replication (BRR) methods.

In Chap. 11, Poisson sampling and its various modifications like collocated sampling are dealt with. The role of Permanent Random Numbers (PRN) in business surveys is briefly covered.

In Chap. 12, Network sampling and Adaptive sampling techniques and their uses with practical applications are illustratively discussed.

In Chap. 13, how sample-sizes are to be settled in varying probability sampling by using simple models are discussed. Allied problems when rather than Direct Response (DR) survey data, Randomized Response (RR) data are also required to be attended to.

Chapter 14 considers multiple frame data consideration and utilizes conditional inference techniques.

In Chap. 15, we consider tests of goodness of fit, test of independence homogeneity and regression and categorical data analyses when samples are chosen with unequal selection probabilities.

Lastly, we may mention that every chapter is concluding with a lessons and exercises section, expected to be worked out by the readers to test how far they are equipped on absorbing the intakes thus far consumed. Some results of our own case studies are also cited whenever considered appropriate for helping the reader's capacity for digestion.

References

Bolfarine, H. & Zacks, S. (1992). *Prediction Theory for Finite Populations*, New York, NY: Springer.

Chaudhuri, A., Christofides, T. C. & Rao, C. R., (2016). *Handbook of Statistics, Data Gathering, Analysis and Protection of Privacy Through Randomized Response Techniques: Qualitative and Quantitative Human Traits*, 34, NL: Elsevier.

Rao, J. N. K. (2003). *Small Area Estimation*, NY, USA: Wiley Interscience.

Rao, J. N. K. & Molina, J. (2015). *Small Area Estimation*, 2nd edition, N.Y., USA: John Wiley & Sons Inc.

Salehi, M. M. & Seber, G. A. F. (2013). *Adaptive sampling designs. Inference for sparse and clustered population*, Heidelberg: Springer.

Särndal, C. E., Swensson, B. E. & Wretman, J. H. (1992). *Model Assisted Survey Sampling*, N.Y.: Springer Verlag.

Thompson, S. K. (1992). *Sampling*, NY, USA: Wiley.

Thompson, M. E. (1997). *Theory of sample surveys*, London: Chapman & Hall.

Thompson, S. K. & Seber, G. A. F. (1996). *Adaptive sampling*, NY, USA: Wiley.

Valliant, R., Dorfman, A. & Royall, R. M. (2000), *Finite Population Sampling and Inference: A Prediction Approach*. New York: John Wiley.

Wu, C. & Thompson, M. E. (2020). *Sampling theory and practice*, Springer.

Contents

About the Authors

Arijit Chaudhuri is an honorary visiting professor at the Applied Statistics Unit at the Indian Statistical Institute (ISI), Kolkata, India, since 1st September 2005. Professor Chaudhuri holds a Ph.D. in Statistics in the area of sample surveys from the University of Calcutta, Kolkata, India, from where he also has graduated. He worked as a postdoctoral researcher for two years at the University of Sydney (1973–1975), Australia. He retired as a professor from the ISI, Kolkata, India, on 31st August 2002, where he then continued to work as a CSIR emeritus scientist for three years up to 31st August 2005. His areas of research include mean square error estimation in multi-stage sampling, analytical study of complex surveys, randomized response surveys, and small area estimation. In the year 2000, he was elected as the President of the Section of Statistics for the Indian Science Congress and worked for the Government of West Bengal for 12 years as the Committee Chairman for the improvement of crop statistics. He has also worked with the Government of India to apply sophisticated methods of small area estimation in improving state and union territory level estimates of various parameters of national importance. He has worked on various global assignments upon invitation, including visiting professorships at several universities and statistical offices in the USA, Canada, England, Germany, the Netherlands, Sweden, Israel, Cyprus, Turkey, Cuba, and South Africa, from 1979 to 2009. He has successfully mentored 10 Ph.D. students and published more than 150 research papers in peer-reviewed journals, a few of them jointly with his students and colleagues. He is the author of 11 books on survey sampling.

Sanghamitra Pal is an assistant professor at the Department of Statistics, West Bengal State University (WBSU), India, since 2009. She completed her Ph.D. in Statistics from the Indian Statistical institute (ISI), Kolkata, in 2004. Earlier, she served as a research scientist from 2001 to 2009 at River Research Institute (RRI), the Government of West Bengal, India. She also worked as a research associate and a visiting scientist at the Applied Statistics Unit, ISI, Kolkata, during January 2005 to July 2006. She is guiding two Ph.D. students in the area of sample surveys at WBSU and has published research articles in several reputed journals. She organized ab invited session on the adaptive cluster sampling at the International Statistical Institute Conference in Durban, South Africa (2009); a sample survey session at the Indian Science Congress (2012); and presented papers at the international conferences in South Africa, New Zealand, Germany, France, Thailand, Singapore, as well as India. She was involved in the professional attachment training (PAT) program of sample survey with ICAR-NAARM, Hyderabad, India. Besides teaching and research, Dr. Pal is working as a nodal officer for the All India Survey of Higher Education cell of WBSU.

Chapter 1
Meaning and Purpose of Survey Sampling

1.1 Definitions

Whenever we talk about sampling, there must be an underlying 'population' or a 'totality' of which a part is a 'sample'.

Symbolically, a population is $U = (1, \ldots, i, \ldots, N)$ consisting of a number of labels denoted by i which stands for one of a total number of elements or units which together constitute a totality called a population of a finite number of N distinct entities each standing for an 'identifiable' object. When described in detail the simple i may denote a household in an actual residential building, say, in a particular street in a specified locality in a city or a town or even in a village, having in s, say, a postal address. The population is regarded as finite as N here is a finite number, though possibly sizeably large as, say, one million or quite small too, as little as 2.

On this population U are defined real variables like, say, y, x, z, w, etc. with y_i, x_i, z_i, w_i as their respective values taken on the unit i of U yielding the vectors

$$\underline{Y} = (y_1, \ldots, y_i, \ldots, y_N),$$
$$\underline{X} = (x_1, \ldots, x_i, \ldots, x_N),$$
$$\underline{Z} = (z_1, \cdots, z_i, \cdots, z_N) \text{ and}$$
$$\underline{W} = (w_1, \ldots, w_i, \ldots, w_N)$$

having, respectively, the total and mean values as

$$Y = \sum_1^N y_i, \bar{Y} = \frac{Y}{N}, X = \sum_1^N x_i, \bar{X} = \frac{X}{N}, \text{ etc.}$$

© The Author(s), under exclusive license to Springer Nature Singapore Pte Ltd. 2022   1
A. Chaudhuri and S. Pal, *A Comprehensive Textbook on Sample Surveys*, Indian Statistical Institute Series, https://doi.org/10.1007/978-981-19-1418-8_1

The numbers $\frac{1}{N}\sum_{1}^{N}(y_i - \bar{Y})^2$, $\frac{1}{N}\sum_{1}^{N}(x_i - \bar{X})^2$ are their respective variances and

$\frac{1}{N}\sum_{1}^{N}(y_i - \bar{Y})(x_i - \bar{X})$ is called the covariance between y and x; also,

$$\frac{\sum_{1}^{N}(y_i - \bar{Y})(x_i - \bar{X})}{\sqrt{\sum_{1}^{N}(y_i - \bar{Y})^2}\sqrt{\sum_{1}^{N}(x_i - \bar{X})^2}}$$

is called the correlation coefficient between y and x.

Any such quantities involving all the population values concerning variables are called 'Parameters'. By dint of observations or surveys, these quantities are required to be assessed either by determining all the values involved by a complete enumeration also called a 'census' or estimated by dint of taking from U a suitably defined 'sample' or a 'part' of the population and ascertaining the variate-values involved in the sample elements actually surveyed.

Thus, $s = (i_1, \ldots, i_j, \ldots, i_n)$ consisting of 1st,..., jth,..., nth elements or units in U taken for observation of the variate-values is called a sample of a size n and taking n as an integer number less than N . Again, $d = ((i_1, y_{i1}), \ldots, (i_j, y_{ij}), \ldots, (i_n, y_{in}))$ is called 'survey data' relating to a variable y for which the sampled values are $y_{i1}, \ldots, y_{ij}, \ldots, y_{in}$ for the sample s as above, provided the sample is chosen and y's as above are observed; if not yet observed, d is 'observable data' if s above is sampled or chosen for observation of y-values for the units in s. The observable 'sample data' may also be denoted as

$$d = (s, y_i | i \in s).$$

On choosing or drawing such a sample, observing such a d is called a 'sample survey'.

Any function of d, say, t with $t = t(d) = t(s, \underline{Y}|$ with no $y_i, i \notin s)$ is called a 'statistic'. If not a census is taken but only the sample survey data d are used to make an assessment of a parameter θ relating to $\underline{Y}$ one says that d is utilized employing a statistic $t = t(d)$, then it is a case of estimation and the statistic $t = t(d)$ is then called an 'Estimator'. The difference

$$(t - \theta)$$

though not known is called the error in t being employed as an estimator for the parameter θ.

A sample s from U is usually drawn by assigning to it a selection probability $p(s)$ on employing a probability measure p to all possible samples like s that one may contemplate to choose from U.

This probability measure p is called a 'Sampling Design' or just a 'Design'. Of course, $0 \leq p(s) \leq 1$ for every s and sum of $p(s)$ over all possible s is unity. The expected value or the expectation of t is then defined as

$$E(t) = \sum_s p(s)t(s, \underline{Y}) = \sum_s p(s)t(d)$$

and $E(t - \theta)$ is called the bias denoted as $B(t)$ of t in estimating θ. If $E(t) = \theta$ for every $\underline{Y}$, then t is regarded as an unbiased estimator for θ. The quantity $M = M(t) = E(t - \theta)^2$ is called the Mean Square Error (MSE) of t as an estimator for θ.

The quantity $V(t) = \sigma^2(t) = E(t - E(t))^2$ is called the variance of t. It follows that $M(t) = V(t) + B^2(t)$ and $\sigma(t) = +\sqrt{V(t)}$ is called the standard deviation or standard error of the estimator t for θ.

Intuitively, a design p and an estimator t for a parameter θ should be so judiciously chosen that the error $(t - \theta)$ should be controlled in magnitude in such a way that the bias $B(t)$ and Mean Square Error (MSE), namely $M(t) = V(t) + B^2(t)$ or at least $|B(t)|$ and $M(t)$ should both be quite small in magnitudes. This is because for every sample one cannot hope the error to be kept small in a desirable way. Thus, we can only choose a design p and an estimator t, the pair (p, t), called a strategy in estimation, in such a way that the 'error' may be desirably kept under check in the sense of average but not 'the observed sample-wise in terms of the realized value of t', so that (i) $|B(t)|$ as well as (ii) $M(t)$ may both be kept under check. Or more simply (i) $B(t)$ may be kept at nil and (ii) $V(t)$ at a minimum possible value.

1.2 Random Sampling

One simple example of a sample called a Simple Random Sample With Replacement (SRSWR) is the one obtained by taking first a unit, say, i_1 of U with a probability $\frac{1}{N}$ and successively units $i_2, \ldots, i_n$ in the same way each with probability $\frac{1}{N}$ from U without caring for re-appearance of any unit any number of times. To do so, the selection procedure is to use a Table of Random Numbers and repeatedly and independently draw one of the labels in U at random keeping U intact during each of the n draws in succession; effectively every time a label is selected from U putting it back to the population before making the next draw. A random number is a number in the infinite sequence of the digits $0, 1, 2, \ldots, 9$ one after another so arranged that anywhere in the sequence a part with K consecutive integers or 0 occurs with a common relative frequency of $\frac{1}{10^K}$, with K, a finite integer number, say, $1 \leq K \leq 8$. A big series like this is called a Table of Random Numbers. Chaudhuri (2010) is a source of which pages 24–26 may be consulted to know a little more of it. A sample

$s = (i_1, \ldots, i_n)$ has then the selection probability $\frac{1}{N^n}$ and every unit i_j in this s has the selection probability $\frac{1}{N}$; also, the pair of units (i_j, i_k) has the selection probability $\frac{1}{N^2}$ and so on.

In a slightly different situation when (i) the units $(i_1, i_2, \ldots, i_j, \ldots, i_n)$ are each a distinct unit of U and (ii) the order of appearances of the labels $i_1, \ldots, i_j, \ldots, i_n$ of U is ignored, we refer to and denote the sample by $s^* = \{i_1, i_2, \ldots, i_j, \ldots, i_n\}$. The selection probability of this s^* is then

$$p(s^*) = \left(\frac{1}{N} \frac{1}{N-1} \frac{1}{N-2} \cdots \frac{1}{(N-n+1)} \right) n! = \frac{(N-n)!n!}{N!} = \frac{1}{\left(N_{C_n} \right)}.$$

This is because on the 1st draw, a unit i_1 is chosen 'at random' with probability $\frac{1}{N}$; on the next draw, the chosen unit i_1 on the 1st draw is set aside and so from the remaining $(N-1)$ units of U one is selected randomly with probability $\frac{1}{N-1}$; thus, the unit i_1 is not returned to the population; that is, the selection is 'without replacement'. Thus, on the nth draw the unit i_n which is different from the previously selected units $i_1, i_2, \ldots, i_{n-1}$, the selection probability assigned to it is $\frac{1}{N-1} \cdots \frac{1}{(N-n+1)}$.

Thus, for every ordered sample, $(i_1, i_2, \ldots, i_n)$, the selection probability is

$$\frac{1}{N} \frac{1}{N-1} \cdots \frac{1}{N-n+1} = \frac{(N-n)!}{N!}$$

So, for the unordered sample s^*, the selection probability is

$$p(s^*) = \frac{n!(N-n)!}{N!} = \frac{1}{\binom{N}{n}}$$

because in s^* there are $n!$ terms for each ordered label in it, n in number.

This method of taking an unordered sample of n units of U, each distinct from one another, on each of n draws choosing one unit with equal probability from among the units remaining in U after the n successive draws without replacement is called Simple Random Sampling Without Replacement (SRSWOR)—clearly from U altogether $\binom{N}{n}$ samples may be formed each with n labels, each distinct from every other with the order of the units in the sample being disregarded. Since each of them has a common selection probability $\frac{1}{\binom{N}{n}}$, this SRSWOR is also a case of equal probability sampling. Hence, it is 'Simple Random' sampling though from one draw to another, the selection probability changes as $\frac{1}{N}, \frac{1}{N-1}, \frac{1}{N-2}, \ldots, \frac{1}{N-n+1}$.

1.3 Estimation Preliminaries

For SRSWR in n draws as an estimator for the population mean $\bar{Y}$, it is intuitive to take the sample mean

$$\bar{y} = \frac{1}{n} \sum_{r=1}^{n} y_r$$

writing y_r for the value of y for the unit of U chosen on the rth draw, $r = 1, 2, \ldots, n$.

Since $\text{Prob}(y_r = y_i) = \text{Prob}(r\text{th draw yields unit } i)$ for every unit i of N and every draw r, it follows that

$$E(y_r) = \sum_{i=1}^{N} y_i \text{Prob}(y_r = y_i)$$

$$= \frac{1}{N} \sum_{1}^{N} y_i = \bar{Y} \text{ for every } r,$$

one gets

$$E(\bar{y}) = \frac{1}{n} \sum_{r=1}^{n} (\bar{Y}) = \bar{Y}.$$

So, $\bar{y}$ is an unbiased estimator of $\bar{Y}$.

Again,
$$V(y_r) = E\left(y_r - E(y_r)\right)^2 = E(y_r - \bar{Y})^2$$

$$= \sum_{i=1}^{n} (y_i - \bar{Y})^2 \text{ Prob}(y_r = y_i)$$

$$= \frac{1}{N} \sum_{i=1}^{n} (y_i - \bar{Y})^2 = \sigma^2, \text{ say.}$$

But covariance between y_r and $y_{r'}$ for any $r \neq r'$ is

$$\text{Cov}(y_r, y_{r'}) = E(y_r - \bar{Y})(y_{r'} - \bar{Y})$$
$$= E(y_r - \bar{Y})E(y_{r'} - Y) = 0 \ \forall \ r \neq r'$$

because y_r and $y_{r'}$ for every $r \neq r'$ are independent because the draws are independently executed. So, $V(\bar{y}) = \frac{\sigma^2}{n}$. This is also a parameter. So, one may find an estimator for it also. Let us consider

$$s^2 = \frac{1}{(n-1)} \sum_{r=1}^{n} (y_r - \bar{y})^2.$$

Then,

$$\sum_{r=1}^{n} (y_r - \bar{y})^2 = \sum_{r=1}^{n} y_r^2 - n(\bar{y})^2.$$

So,

$$E(s^2) = \frac{1}{(n-1)} \left[\sum_{r=1}^{n} \{V(y_r) + (\bar{Y})^2\} - n \{V(\bar{y}) + (\bar{Y})^2\} \right]$$

$$= \frac{1}{(n-1)} \left[n\sigma^2 + n(\bar{Y})^2 - n \left(\frac{\sigma^2}{n} \right) - n(\bar{Y})^2 \right]$$

$$= \sigma^2.$$

So, $E\left(\frac{s^2}{n} \right) = \frac{\sigma^2}{n}$, i.e., $\frac{s^2}{n}$ is an unbiased estimator for $V(\bar{y})$.

When the unordered sample $s^* = \{i_1, \ldots, i_j, \ldots, i_n\}$ of n distinct units is selected from $U = (1, \ldots, N)$ by SRSWOR method with $p(s^*) = \frac{1}{\binom{N}{n}}$ and a survey is implemented gathering the sample-values $y_{i1}, \ldots, y_{in}$ of y for the labels in s^*, then one may calculate the sample mean

$$\bar{y} = \frac{1}{n} \sum_{i \in s^*} y_i.$$

Then, its expectation is

$$E(\bar{y}) = \frac{1}{\binom{N}{n}} \sum_{s^*} \left(\frac{1}{n} \sum_{i \in s^*} y_i \right)$$

denoting by $\sum_{s^*}$ the sum over all possible samples of the form s^*.

Then,

$$E(\bar{y}) = \frac{1}{n} \frac{1}{\binom{N}{n}} \sum_{i=1}^{N} y_i \left(\sum_{s^* \ni i} 1 \right)$$

Here, $\sum\limits_{s^* \ni i} 1$ means the total number of samples of the type s^* containing the single

unit i of U. This number, clearly, is $\binom{N-1}{n-1}$ because besides the unit i in U there

are $(N-1)$ other units out of which $(n-1)$ distinct units ignoring the ordering of appearance of the units are to be chosen to produce a sample like s^* of $(n-1)$ units omitting in it the unit i of U.

$$\text{So,}\quad E(\bar{y}) = \frac{1}{n}\frac{1}{\binom{N}{n}}\binom{N-1}{n-1}\sum_{1}^{N} y_i$$

$$= \frac{1}{n}\frac{(N-1)!}{(n-1)!}\frac{1}{(N-n)!}\frac{n!(N-n)!}{N!}Y$$

$$= \frac{Y}{N} = \bar{Y}.$$

Thus, $\bar{y}$ is unbiased for $\bar{Y}$ in SRSWOR. Let us next calculate the

$$V(\bar{y}) = E(\bar{y} - \bar{Y})^2 = \frac{1}{n^2}E\left[\sum_{i \in s^*}(y_i - \bar{Y})\right]^2$$

$$= \frac{1}{n^2}\frac{1}{\binom{N}{n}}\sum_{s^*}\left[\sum_{i \in s^*}(y_i - \bar{Y})\right]^2$$

$$= \frac{1}{n^2}\frac{1}{\binom{N}{n}}\sum_{s^*}\left[\sum_{i \in s^*}(y_i - \bar{Y})^2 + \sum\sum_{i \neq j \in s^*}(y_i - \bar{Y})(y_j - \bar{Y})\right]$$

$$= \frac{1}{n^2}\frac{1}{\binom{N}{n}}\left[\sum_{i=1}^{N}(y_i - \bar{Y})^2\sum_{s^* \ni i}1 + \sum\sum_{i \neq j}(y_i - \bar{Y})(y_j - \bar{Y})\sum_{s^* \ni i,j}\right].$$

Here, $\sum\limits_{s^* \ni i,j}$ is the number of samples of the type s^* which contains the distinct

unordered pair of units $\{i, j\}$ and this number is $\binom{N-2}{n-2}$.

$$\text{So,} \quad V(\bar{y}) = \frac{1}{n^2} \frac{1}{\binom{N}{n}} \left[\binom{N-1}{n-1} \sum_1^N (y_i - \bar{Y})^2 + \binom{N-2}{n-2} \sum_i^N \sum_{i \neq j}^N (y_i - \bar{Y})(y_j - \bar{Y}) \right]$$

$$= \frac{1}{n^2} \frac{(N-1)!}{(n-1)!} \frac{1}{(N-n)!} \frac{n!(N-n)!}{N!} \sum_1^N (y_i - \bar{Y})^2$$

$$- \frac{1}{n^2} \frac{(N-2)!}{(n-2)!} \frac{n!}{(N-n)!} \frac{(N-n)!}{N!} \sum_1^N (y_i - \bar{Y})^2$$

because

$$\sum_i \sum_{i \neq j} (y_i - \bar{Y})(y_j - \bar{Y}) = \left[\sum_1^N (y_i - \bar{Y}) \right]^2 - \sum_1^N (y_j - \bar{Y})^2$$

$$= - \sum_1^N (y_i - \bar{Y})^2$$

This simplifies to

$$V(\bar{y}) = \frac{N-n}{nN} \frac{1}{N-1} \sum_1^N (y_i - \bar{Y})^2 = \frac{N-n}{Nn} S^2$$

writing $S^2 = \frac{1}{N-1} \sum_1^N (y_i - \bar{Y})^2$ which equals $\frac{N\sigma^2}{N-1}$. In S^2, it is instructive to use $(N-1)$ as the divisor as it is the number of degrees of freedom $(N-1)$ carried by $\sum_1^N (y_i - \bar{Y})^2$ which is subject to the single constraint, namely

$$\sum_1^N (y_i - \bar{Y}) = 0.$$

This $V(\bar{y})$ is also a parameter and so it is worth estimating.

Let

$$s^2 = \frac{1}{(n-1)} \sum_{i \in s^*} (y_i - \bar{y})^2$$

Now

$$\sum_{i \in s^*} (y_i - \bar{y})^2 = \sum_{i \in s^*} \left[(y_i - \bar{Y}) - (\bar{y} - \bar{Y}) \right]^2$$

$$= \sum_{i \in s^*} (y_i - \bar{Y})^2 - n(\bar{y} - \bar{Y})^2$$

So,
$$E \sum_{i \in s^*} (y_i - \bar{y})^2 = \frac{1}{\binom{N}{n}} \sum_{1}^{N} (y_i - \bar{Y})^2 \left(\sum_{s^* \ni i} 1 \right) - n V(\bar{y})$$

$$= \frac{\binom{N-1}{n-1}}{\binom{N}{n}} \sum_{1}^{N} (y_i - \bar{Y})^2 - n \frac{N-n}{Nn} s^2$$

$$= \left[\frac{n(N-1)}{N} - \frac{(N-n)}{N} \right] s^2$$

$$= (n-1) S^2$$

So,
$$E(s^2) = S^2$$

and $\frac{N-n}{Nn} s^2$ is an unbiased estimator of $V(\bar{y})$.

Since $\bar{Y}$ is a real number and so is $\bar{y}$, both represented by a single point on the real line $(-\infty, +\infty)$, such an estimator is called a 'Point Estimator'. Its performance characteristics are its expectation and bias, Mean Square Error (MSE) or variance in case it is unbiased for the parameter it intends to estimate and an estimator for its MSE or variance.

For an estimator t of a parameter θ, its MSE is

$$M = E(t - \theta)^2 = \sum_s p(s)(t(s, \underline{Y}) - \theta)^2$$

$$= \sum_{s \,:\, |t(s, \underline{Y}) - \theta| \geq K} p(s)(t(s, \underline{Y}) - \theta)^2 + \sum_{s \,:\, |t(s, \underline{Y}) - \theta| < K} p(s)(t(s, \underline{Y}) - \theta)^2$$

for a positive number K.

Then,
$$M \geq K^2 \mathrm{Prob}[|t(s, \underline{Y}) - \theta| \geq K]$$

or
$$\mathrm{Prob}[|t(s, \underline{Y}) - \theta| \geq K] \leq \frac{M}{K^2}$$

or
$$\mathrm{Prob}[|t(s, \underline{Y}) - \theta| \leq K] \geq 1 - \frac{M}{K^2}$$

$$= 1 - \frac{V + B^2}{K^2}$$

$$= 1 - \frac{\sigma^2 + B^2}{K^2}$$

This is known as Chebyshev's inequality. On taking $K = \lambda\sigma$ for a positive number λ, it follows that

$$\text{Prob}\left[|t(s, \underline{Y}) - \theta| \leq \lambda\sigma\right] \geq \left(1 - \frac{1}{\lambda^2}\right) - \frac{1}{\lambda^2}\left(\frac{|B|}{\sigma}\right)^2$$

$$\text{Prob}\left[t(s, \underline{Y}) - \lambda\sigma \leq \theta \leq t(s, \underline{Y}) + \lambda\sigma\right] \geq \left(1 - \frac{1}{\lambda^2}\right) - \left(\frac{|B|}{\sigma}\right)^2 \frac{1}{\lambda^2}.$$

From this, (Neyman, 1934) concluded that the 'Random Interval' $\left(t(s, \underline{Y}) - \lambda\sigma, t(s, \underline{Y}) + \lambda\sigma\right) = CI$, say, contains the fixed but unknown value of θ with at least a probability

$$\left(1 - \frac{1}{\lambda^2}\right) - \frac{1}{\lambda^2}\left(\frac{|B|}{\sigma}\right)^2.$$

This interval CI is, therefore, called a 'Confidential Interval' for θ with the 'Confidence Coefficient' (CC) at least as high as $\left(1 - \frac{1}{\lambda^2}\right) - \frac{1}{\lambda^2}\left(\frac{|B|}{\sigma}\right)^2$. If t is unbiased for θ, then the CI $(t \pm \lambda\sigma)$ covers θ with a CC of magnitude at least $\left(1 - \frac{1}{\lambda^2}\right)$, no matter whatever may be the magnitudes of the co-ordinates of the vector $\underline{Y} = (y_1, \ldots, y_i, \ldots, y_N)$. For example, taking $\lambda = 3$, one may claim that $(t \pm 3\sigma)$ has at least a CC of $\frac{8}{9}$ which is approximately about 89%.

Thus, in case an SRSWR is taken in n draws, then the CI $\left(\bar{y} \pm 3\frac{\sigma}{\sqrt{n}}\right)$ for $\bar{Y}$ has the CC of at least 89%.

Similarly, for an SRSWOR in n draws, the CI $\left(\bar{y} \pm 3S\sqrt{\left(\frac{1}{n} - \frac{1}{N}\right)}\right)$ has a CC of at least 89%.

In cases of SRSWR and SRSWOR, this Chebyshev's inequality may be very effectively utilized in 'specifying the sample-size or the numbers of draws to be recommended' as follows.

Suppose one needs to unbiasedly estimate the finite population mean by employing the sample mean such that the error $(\bar{y} - \bar{Y})$ be so controlled that $|\bar{y} - \bar{Y}|$ should not exceed a fraction $f(0 < f < 1)$ of the unknown mean $\bar{Y}$ with a very high probability. Thus, let it be required that

$$\text{Prob}[|\bar{y} - \bar{Y}| \leq f\bar{Y}] \geq 1 - \alpha$$

with $\alpha(0 < \alpha < 1)$ quite small, say equal to 0.05.

Let us take $\alpha = \frac{1}{\lambda^2}$ and

$$\lambda\frac{\sigma}{\sqrt{n}} = f\bar{Y} \text{ for SRSWR} \tag{1.3.1}$$

$$\lambda S\sqrt{\left(\frac{1}{n} - \frac{1}{N}\right)} = f\bar{Y} \text{ for SRSWOR.} \tag{1.3.2}$$

Table 1.1 Fixing sample-size in SRSWOR

N (1)	$100f$ (2)	α (3)	CV (4)	n (upward to integer) (5)
100	10	0.05	10	
80	10	0.05	12	
60	12	0.05	15	
50	15	0.05	10	

From (1.3.2), one gets $\left(\frac{1}{n} - \frac{1}{N}\right) = \alpha \left(\frac{100f}{\text{CV}}\right)^2$ on writing CV $= 100\frac{S}{\bar{Y}}$, the coefficient of variation of the values of y_i in $\underline{Y} = (y_1, \cdots, y_i, \ldots, y_N)$. This gives

$$n = \frac{N}{1 + N\alpha \left(\frac{100f}{\text{CV}}\right)^2} \qquad (1.3.3)$$

as a solution by Chebyshev's inequality a rule to fix n in an SRSWOR to unbiasedly estimate $\bar{Y}$.

This rule (1.3.3) permits a tabulation scheme to lead to Table 1.1.

A similar table follows for SRSWR too.

If you are not inclined to anticipate possible values of CV $= 100\frac{S}{\bar{Y}} = 100\frac{\sigma}{\bar{Y}}\sqrt{\frac{N}{N-1}}$, then also it is possible to scientifically fix sample-sizes in at least two alternative ways, with or without using Chebyshev's inequality.

First, in the cases of SRSWR and SRSWOR, one may postulate that $\frac{\bar{y}-\bar{Y}}{\sigma}$ or $\frac{\bar{y}-\bar{Y}}{S}$, respectively, is distributed as a standardized normal deviate τ which has the distribution $N(0, 1)$ with the probability density function

$$f(\tau) = \frac{1}{\sqrt{2\pi}} e^{\frac{\tau^2}{2}}.$$

Thus, it is well-known that

$$\text{Prob}[\tau \geq 1.96] = 0.05$$

or

$$\text{Prob}[\tau \leq -1.96] = 0.95$$

So, in case of SRSWR,

$$\text{Prob}[|\bar{y} - \bar{Y}| \leq f\bar{Y}]$$

$$= \text{Prob}[|\tau| \leq \frac{f\bar{Y}}{\sigma}\sqrt{n}] = 0.95$$

if $\frac{100f}{\text{CV}}\sqrt{n} = 1.96$ or $n = (1.96)^2 \frac{(\text{CV})^2}{(100f)^2}$ since CV $= 100\frac{\sigma}{\bar{Y}}$.

So one may tabulate

N	$100f$	CV	n (correct upward to integer)
60	10	10	
100	5	12	

In case of SRSWOR, since the CV is $100\frac{S}{\bar{Y}}$,

$$\text{Prob}\left[|\bar{y} - \bar{Y}| \le f\bar{Y}\right]$$

$$= \text{Prob}\left[|\tau| \le \frac{f\bar{Y}\sqrt{n}}{S\sqrt{\frac{N-1}{N}}}\right] = 0.95$$

if

$$\frac{100f}{\text{CV}}\frac{\sqrt{n}}{\sqrt{\frac{N-1}{N}}} = 1.96.$$

So, one may tabulate

$$n = \frac{N}{1 + \frac{N(100f^2)}{(1.96)^2(\text{CV})^2}}$$

using

N	$100f$	CV	n (correct upward to integer)
60	10	10	
80	10	12	
100	20	10	

1.4   Requirements of a Good Sample

With the knowledge acquired from what precedes, a good sample should have a specified probability of selection and an estimator to be based on such a probability sample should have a small (or nil) numerical value of its bias vis-a-vis a parameter it is to seek to estimate with a small magnitude of its variance or the Mean Square Error. More will you gather later as we proceed further.

1.5 Lessons and Exercises

Lessons

(i) Given $N=8$, $n=3$ and the samples by SRSWR as

$$s_1 = (3, 5, 7)$$
$$s_2 = (7, 2, 1)$$
$$s_3 = (5, 4, 5).$$

Write down $p(s_1)$, $p(s_2)$, $p(s_3)$.

(ii) For (i) let $y_1 = 20$, $y_2 = 4$, $y_3 = -3$, $y_4 = 6.20$, $y_5 = 0$, $y_6 = 17$, $y_7 = \frac{1}{2}$ and $y_8 = 4$.

Write down the detailed survey data for the three samples in (i). Write down the values of $\bar{y}$ for the three samples in (i) and evaluate an unbiased estimate for $V(\bar{y})$ based on s_3 above.

(iii) For the survey population $U = (1, 2, 3, \ldots, 11)$, you are required to draw an SRSWOR of size 4.

Let $\underline{Y} = (10, -4, 0, 0, 9, 19, 17, -2, 1, 1, 1)$. For the sample you draw evaluate the sample mean and an unbiased estimate of the variance of the sample mean. Also, find an appropriate sample-size if the error in estimating the population mean by the sample mean is not to exceed in magnitude the value (a) 4.2 with a 95% CC chance applying Chebyshev's inequality or by (b) normality assumption for the standardized error with a CC of 95% with the numerical error not exceeding 10% of the unknowable population mean.

(iv) If (t_1, p_1) and (t_2, p_2) denote two strategies to estimate a common parameter θ, then $V_{p_1}(t_1)$ and $V_{p_2}(t_2)$ will denote the two variances of the two respective estimators t_1, t_2 based, respectively, on the designs p_1 and p_2. Then

$$E = 100 \frac{V_{p_2}(t_2)}{V_{p_1}(t_1)}$$

is called the efficiency of (t_1, p_1) relative to (t_2, p_2). Since for a good estimator a lower variance is desirable, t_1 will be better than t_2 if $E \geq 100\%$. If t_1 and t_2 are same but p_1 and p_2 are different, then p_1 will be better than p_2 in case $E \geq 100\%$.

Discuss how SRSWOR is better than SRSWR if both involve the same (N, n) and $n \geq 2$ and the sample mean in both cases is used to estimate the population mean.

Exercises

(a) Suppose $\pi_i = \sum_{s \ni i} p(s)$ denotes the 'Inclusion probability' of the unit i of a population in a sample chosen according to a sampling design p. Also, suppose

$\pi_{ij} = \sum_{s \ni i,j} p(s)$ is the 'Inclusion probability' of the pair of units (i, j) of a population in a sample chosen according to a design p. In case $j = i$, we shall write
$\pi_{ii} = \pi_i$.

Calculate π_i, π_{ij} separately for SRSWOR and SRSWR.

For SRSWOR with (N, n),

$$\pi_i = \frac{1}{\binom{N}{n}} \sum_{s \ni i} 1 = \frac{\binom{N-1}{n-1}}{\binom{N}{n}} = \frac{n}{N} \; \forall i.$$

$$\pi_{ij} = \frac{1}{\binom{N}{n}} \sum_{s \ni i,j} 1 = \frac{\binom{N-2}{n-2}}{\binom{N}{n}} = \frac{n(n-1)}{N((N-1)} \; \forall i, j (i \neq j).$$

To calculate π_i, π_{ij} for SRSWR, it will be convenient first to calculate $\pi_i^C =$ the
complement of π_i, i.e., the exclusion-probability of i in a sample chosen from
the population by SRSWR, which is

$$\pi_i^C = 1 - \pi_i = \sum_{s \not\ni i} p(s) = \left(\frac{N-1}{N}\right)^n.$$

So, $\pi_i = 1 - \left(\frac{N-1}{N}\right)^n$ for SRSWR with (N, n).

Let A_i denote the event that the unit i is 'included in' and A_i^C denote i is
'excluded from' a sample chosen by SRSWR in case (N, n). Then, $(A_i \cup A_j)$
denotes either i or j or i and j are both included in an SRSWR in case of (N, n).
Also, $A_i^C \cap A_j^C$ denote that both ith and jth units are both excluded from an
SRSWR in case (N, n).
'De Morgan's Law' says

$$\left(A_i \cup A_j\right)^C = \left(A_i^C \cap A_j^C\right) \text{ and}$$
$$\left(A_i \cap A_j\right)^C = \left(A_i^C \cup A_j^C\right).$$

$$\text{So,}\quad \pi_{ij} = \text{Prob}\left(A_i \cap A_j\right) = 1 - P\left(A_i \cap A_j\right)^C$$
$$= 1 - \left[P\left(A_i^C \cup A_j^C\right)\right]$$
$$= 1 - \left[P\left(A_i^C\right) + P\left(A_j^C\right) - P\left(A_i^C \cap A_j^C\right)\right]$$
$$= 1 - \left[1 - \left(\frac{N-1}{N}\right)^n + 1 - \left(\frac{N-1}{N}\right)^n - \left(\frac{N-2}{N}\right)^n\right]$$
$$= 2\left(\frac{N-1}{N}\right)^n + \left(\frac{N-2}{N}\right)^n - 1.$$

The above result is of interest because of the following consequences.

Let
$$I_{si} = 1 \text{ if } s \ni i$$
$$= 0 \text{ if } s \not\ni i.$$

So,
$$\pi_i = \sum_{s \ni i} p(s) = \sum_{s} p(s) I_{si} = E(I_{si})$$

and
$$\sum_{i=1}^{N} \pi_i = \sum_{s} p(s)\left(\sum_{i=1}^{N} I_{si}\right)$$
$$= \sum_{s} p(s)\nu(s)$$

writing $\nu(s) = \sum_{i=1}^{N} I_{si}$ = Number of distinct units in a sample s.

So,
$$\sum_{i=1}^{N} \pi_i = E(\nu(s))$$
$$= \text{Average number of distinct units in the samples} = \nu, \text{ say.}$$

So, in case of (N, n), an SRSWR contains on an average the number of distinct units which is

$$\nu = \sum_{i=1}^{N}\left[1 - \left(\frac{N-1}{N}\right)^n\right]$$
$$= N\left[1 - \left(\frac{N-1}{N}\right)^n\right].$$

This is approximately equal to

$$\nu = N - N \left[1 - \frac{n}{N} + \frac{n(n-1)}{2N^2} \right]$$

$$= n - \frac{n(n-1)}{2N}$$

$$< n.$$

For example, if $N = 10$, $n = 4$, then $\nu = 4 - \frac{4 \times 3}{2 \times 10} = 3.4$. If $N = 50$, $n = 18$, ν equals 14.94.

(b) Given the sample-size n for an SRSWOR with (1) the population size N suppressed but with (2) the selected labels given.

Freund (1992) has given a method to 'unbiasedly estimate N'. He needs the probability distribution of X, the highest label sampled in SRSWOR (N, n) and this is

$$\text{Prob}(X = x) = \frac{\binom{x-1}{n-1}\binom{1}{1}}{\binom{N}{n}} \text{ for } x = n, n+1, \dots, N. \tag{a}$$

So,
$$E(X) = \frac{1}{\binom{N}{n}} \sum_{x=n}^{N} x \binom{x-1}{n-1} = \frac{1}{\binom{N}{n}} \sum_{n}^{N} \frac{x!}{(n-1)!(x-n)!}. \tag{b}$$

From (a) follows

$$\binom{N}{n} = \sum_{x=n}^{N} \frac{(x-1)!}{(n-1)!(x-n)!}.$$

So, (b) gives

$$E(X) = \frac{n}{\binom{N}{n}} \sum_{x=n}^{N} \frac{x!}{n!(x-n)!}$$

$$= \frac{n}{\binom{N}{n}} \sum_{(x+1)=n+1}^{N+1} \frac{(x+1-1)!}{\{(x+1)-(n+1)\}!(n+1-1)!}$$

$$= n \frac{1}{\binom{N}{n}} \binom{N+1}{n+1}$$

$$= \frac{n}{N!} \frac{n!(N-n)!(N+1)!}{(n+1)!(N-n)!}$$

$$= \frac{n(N+1)}{(n+1)}.$$

So, an unbiased estimator of N is

$$\widehat{N} = \frac{(n+1)X}{n} - 1.$$

Find $V(\widehat{N})$ and an unbiased estimator of $V(\widehat{N})$ using the above.

(c) From (a), it is instructive to contemplate using for $\bar{Y}$ an alternative unbiased estimator in case an SRSWR, (N, n) is employed examining the following: We may recall that

$$E(I_{si}) = \pi_i.$$

So, for $Y = \sum_1^N y_i$ we may propose the estimator

$$t_{HT} = \sum_{i=1}^N y_i \frac{I_{si}}{\pi_i} \text{ if } \pi_i > 0 \forall i \in U$$

because $E(t_{HT})$ equals Y.

This estimator was given to us by Narain (1951) and Horvitz and Thompson (1952). We may further observe that

$$V(I_{si}) = E(I_{si})[1 - E(I_{si})]$$
$$= \pi_i - \pi_i^2 = \pi_i(1 - \pi_i)$$

because $I_{si}^2 = I_{si}$.

Also, the covariance between I_{si} and I_{sj} for $i \neq j$ is

$$\text{Cov}(I_{si}, I_{sj}) = E(I_{si} I_{sj}) - \pi_i \pi_j$$
$$= \pi_{ij} - \pi_i \pi_j.$$

$$\text{So,} \quad V(t_{HT}) = E(t_{HT}^2) - [E(t_{HT})]^2$$

$$= \sum_{1}^{N} \frac{y_i^2}{\pi_i} + \sum\sum_{i \neq j} y_i y_j \frac{\pi_{ij}}{\pi_i \pi_j} - Y^2$$

$$= \sum_{1}^{N} y_i^2 \left(\frac{1}{\pi_i} - 1 \right) + \sum\sum_{i \neq j} y_{ij} \left(\frac{\pi_{ij}}{\pi_i \pi_j} - 1 \right).$$

Since $E(I_{sij}) = E(I_{si} I_{sj}) = \pi_{ij}$, it follows that

$$\nu(t_{HT}) = \sum_{i=1}^{N} y_i^2 \left(\frac{1 - \pi_i}{\pi_i} \right) \frac{I_{si}}{\pi_i} + \sum\sum_{i \neq j} y_i y_j \left(\frac{\pi_{ij} - \pi_i \pi_j}{\pi_i \pi_j} \left(\frac{I_{sij}}{\pi_{ij}} \right) \right).$$

This unbiased estimator $\nu(t_{HT})$ is also given to us by Horvitz and Thompson (1952).

Now, if one employs SRSWR (N, n), then as an alternative to $\bar{y}$ one may resort to the HT estimator

$$\bar{y}_{SRSWRHT} = \frac{1}{N} \sum_{i \in s} \frac{y_i}{\pi_i}$$

with π_i given as earlier.

Also, $\nu(\bar{y}_{SRSWRHT}) = \frac{1}{N^2} \nu(t_{HT})$ with π_i, π_{ij} as given earlier.

No case studies worth reporting can be described relevant to what is covered so far in this chapter.

References

Neyman, J. (1934). On the two different aspects of the representative method, the method of stratified sampling and the method of purposive selection. *JRSS, 97*, 558–625.

Freund, J. E. (1992). *Mathematical statistics* (5th ed.). N.J., USA: Prentice Hall Inc. Englewood Cliffs.

Narain, R. D. (1951). On sampling without replacement with varying probabilities. *JISAS, 3*, 169–175.

Horvitz, D. G., & Thompson, D. J. (1952). A generalization of sampling without replacement from a finite universe. *JASA, 47*, 663–689.

Chaudhuri, A. (2010). *Essentials of survey sampling*. New Delhi, India: PHI.

Chapter 2
Inference in Survey Sampling

2.1 Traditional or Design-Based Approach: Its Uses and Limitations

A few alternative approaches in addressing the inference-making in the context of Survey Sampling have by now emerged which we are inclined to take up for narration in this treatise. In the present chapter, we shall treat only the classical or the traditional approach, also called the Design-based approach. This involves considering selection of a sample s from the specified survey population $U = (1, \ldots, i, \ldots, N)$ assigning to it a selection probability $p(s)$, which does not involve the values y_i, $i = 1, \ldots, N$ of the variable of interest y, a parameter θ relating to $\underline{Y} = (y_1, \ldots, y_i, \ldots, y_N)$ being our centre of interest to make an inference about point and/or interval estimation. Once the sample s is selected, it is to be surveyed to gather the associated y-values for y_i for $i \in s$ and on obtaining the survey data $d = \big((i_1, y_{i1}), \ldots, (i_j, y_{ij}), \ldots, (i_n, y_{in})\big)$ if the selected sample is

$$s = (i_1, \ldots, i_j, \ldots, i_n)$$

or any reduced data, say

$$d^* = (s^*, y_i | i \in s^*)$$

with, say, $s^* = \{i_1, \ldots, i_k\}$ for k in s ignoring the order and frequency of the units in s.

Then, only p-based characteristics related to functions of d like $t = t(d) = t(s, \underline{Y})$ involving y_i's only for $i \in s$ but free of y_j for $j \notin s$ though $j \in U$, like expectation $E(t)$, bias $B(t) = E(t - \theta)$, $M(t) = \text{MSE} = E(t - \theta)^2$, $V(t) = E(t - E(t))^2$, $\sigma(t) = +\sqrt{V(t)}$ are considered to study the performance characteristics of t vis-a-vis θ as noted above for example. An estimator t for θ is employed to control the error $(t - \theta)$ and magnitude of the error $|t - \theta|$ so as to keep $E(t - \theta) = B(t)$ or $|B(t)|$ small and the MSE $E(t - \theta)^2 = V(t) + B(t)^2$ small in magnitude. Given a chosen sample s and an estimator t for θ proposed based on

A. Chaudhuri and S. Pal, *A Comprehensive Textbook on Sample Surveys*, Indian Statistical Institute Series, https://doi.org/10.1007/978-981-19-1418-8_2

the data d and the statistic $t = t(s, \underline{Y})$ free of y_j for $j \notin s$, this traditional, classical Design-based approach of inference about θ through t does not aspire to say how good is t or how close is t to θ for the sample surveyed and the data gathered at hand. But the survey statistician is concerned about the average magnitudes of $B(t), |B(t)|, M(t)$ and $\sigma(t)$ or of the coefficients of variation $CV = 100\frac{\sigma(t)}{Y}$. As we noted both for 'point and interval estimation' it is desirable that $|B(t)|$ and $\sigma(t)$ should be numerically small. But later, we shall find that among unbiased estimators, for a finite population total based on plausible designs, no estimator exists with a uniformly smallest variance. Also, though it will be shown that a 'complete class of estimators' may be constructed with the property that for a given estimator outside this class, a better estimator exists within this 'complete' class, but this 'complete' class is too wide to be 'effective'. Because of such prohibitive limitations, the 'Design-based' approach is not adequate to give us an acceptable solution. But this approach is simple to understand and apply and so is an overwhelming implementation in practice. But alternative approaches are necessary and they are forthcoming in plenty and interestingly pursued.

2.2   Sampling Designs and Strategies

We have mentioned already that p is a sampling design which is a probability measure assigning a probability $p(s)$ to a sample s from the population U with which it is to be selected. If an estimator t based on s is a statistic $t = t(s, \underline{Y})$, free of y_j for $j \notin s$ is to be employed to estimate a parameter θ related to $\underline{Y} = (y_1, \ldots, y_i, \ldots, y_N)$, then the combination (p, t) is called a 'strategy'. For example, $(SRSWR(N, n), N\bar{y})$ is a 'strategy' to unbiasedly estimate $Y = N\bar{Y}$ and $(SRSWOR(N, n), N\bar{y})$ is another 'strategy' for the same purpose. This book is intended to illustrate many other such 'strategies' to estimate Y or $\bar{Y}$ as we shall present them in succession indicating their pros and cons in appropriate places.

2.3   Inclusion Probabilities and Their Inter-Relationships

In Sect. 1.2 of Chap. 1, we introduced π_i and π_{ij} as inclusion probabilities, respectively, of a unit i and pair of units i and j of U for a sampling design p. Also, we defined

$$I_{si} = 1, \ \text{if } i \in s$$
$$= 0, \ \text{if } i \notin s$$

and
$$I_{sij} = I_{si}I_{sj}$$
$$= 1 \text{ if both } i \text{ and } j \in s$$
. $$= 0 \text{ if at least one of } i \text{ and } j \text{ is not contained in } s. \quad (2.3.1)$$

We further observed that

$$\sum_{i=1}^{N} \pi_i = \sum_{i=1}^{N}\left(\sum_s p(s)I_{si}\right)$$

$$= \sum_s p(s)\left(\sum_{i=1}^{N} I_{si}\right)$$

$$= \sum_s p(s)v(s)$$

$$= E(v(s))$$

$$= v \quad (2.3.2)$$

denoting by $v(s)$, the number of distinct units in s and v denoting the expected value of $v(s)$.

Now,
$$\sum_{j\neq i}^{N} \pi_{ij} = \sum_{j\neq i}\left(\sum_{s\ni i,j} p(s)\right)$$

$$= \sum_{j\neq i}\left(\sum_s p(s)I_{sij}\right)$$

$$= \sum_s p(s)\left(\sum_{j\neq i} I_{sij}\right)$$

$$= \sum_s p(s)\left(I_{si}\sum_{j\neq i} I_{sj}\right)$$

$$= \sum_s p(s)\left((v(s) - 1)I_{si}\right)$$

$$= \sum_s p(s)v(s)I_{si} - \pi_i$$

$$= \sum_{s\ni i} p(s)v(s) - \pi_i \quad (2.3.3)$$

$$\text{So} \sum_{\substack{i=1 \\ i\neq j}}^{N} \sum_{j=1}^{N} \pi_{ij} = \sum_{s} p(s)v^2(s) - \sum_{1}^{N} \pi_i$$

$$= E\left(v^2(s)\right) - E\left(v(s)\right)$$
$$= V\left(v(s)\right) + E^2\left(v(s) - Ev(s)\right)$$
$$= V\left(v(s)\right) + v(v-1). \tag{2.3.4}$$

If $v(s) = n$, for every s with $p(s) > 0$, (2.3.2) and (2.3.3), respectively, yield

$$\sum \pi_i = n, \sum_{j\neq i} \pi_{ij} = (n-1)\pi_i \text{ and}$$

$$\sum\sum_{i\neq j} = n(n-1).$$

Thus, the inclusion probabilities are not independently assignable non-negative numbers, but are subject to the above "Consistency Conditions".

2.4 Necessary and Sufficient Conditions for Existence of Unbiased Estimators

For a finite population total $Y = \sum_{i=1}^{N} y_i$ an unbiased estimator $t = t(s, \underline{Y})$, free of y_j for $j \notin s$, for Y must satisfy the requirement that

$$Y = \sum_{i=1}^{N} y_i = \sum_{s} p(s)t(s, \underline{Y}) \forall \underline{Y}.$$

Thus we need

$$Y = \sum_{i=1}^{N} y_i = \sum_{s\ni i} p(s)t(s, \underline{Y}) + \sum_{s\not\ni i} p(s)t(s, \underline{Y}).$$

Suppose the design p is such that for an i, $\pi_i = 0$, i.e., $0 = \sum_{s\ni i} p(s) \to p(s) = 0 \forall s$ such that $s \ni i$. Then,

$$Y = \sum_{i=1}^{N} y_i = y_i + \sum_{j\neq i}^{N} y_j = \sum_{s\not\ni i} p(s)t(s, \underline{Y}) \tag{2.4.1}$$

By definition of an estimator $t = t(s, \underline{Y})$, the right-hand side of (2.4.1) is free of y_i and so the equality (2.4.1) cannot hold for every real number y_i. So, if $\pi_i = 0$ for any i in $U = (1, \ldots, i, \ldots, N)$ an unbiased estimator for Y cannot exist. This means a 'Necessary Condition' for existence of an unbiased estimator for a population total is that $\pi_i > 0 \, \forall i$. On the other hand, if $\pi_i > 0 \, \forall i \in U$, then for $Y = \sum_1^N y_i$ an unbiased estimator is available as

$$t_{HT} = \sum_{i=1}^N y_i \left(\frac{I_{si}}{\pi_i} \right) = \sum_{i \in s} \frac{y_i}{\pi_i}.$$

Hence, we have the

Theorem 1 *A necessary and sufficient condition for the existence of an unbiased estimator for a finite population total is*

$$\pi_i > 0 \, \forall i \in U.$$

Notes

(1) Similarly, one may check that a 'Necessary and Sufficient Condition' for the existence of an unbiased estimator for

$$Y^2 = \left(\sum_1^N y_i \right)^2 = \sum_{i=1}^N y_i^2 + \sum \sum_{i \neq j} y_i y_j$$

is that $\pi_{ij} > 0 \, \forall i, j (i \neq j) \in U$.

(2) Observing further that $\pi_{ij} > 0 \, \forall i, j (i \neq j) \in U$, implies $\pi_i > 0 \, \forall i \in U$.

(3) If $E(t) = Y$, then, $V(t) = E(t^2) - Y^2$; hence $t^2 - \widehat{Y^2}$, writing $\widehat{Y^2}$ as an unbiased estimator for Y^2 if such an $\widehat{Y^2}$ exists, is unbiased for $V(t)$; so a necessary and sufficient condition for the existence of an unbiased estimator for the variance of an unbiased estimator for a finite population total is

$$\pi_{ij} > 0 \, \forall i \neq j \text{ with } i, j \in U = (1, \ldots, N)$$

2.5 Linear and Homogeneous Linear Unbiased Estimators

Godambe (1955) defined for a population total $Y = \sum_{i=1}^N y_i$ a class of homogeneous linear unbiased estimators of the form

$$t_b = \sum_{i \in s} y_i b_{si}$$

with b_{si} as a constant free of $\underline{Y} = (y_1, \ldots, y_i, \ldots, y_N)$ subject to the condition

$$\sum_{s \ni i} p(s) b_{si} = 1 \, \forall i \in U. \tag{2.5.1}$$

This restriction is needed because

$$E(t_b) = \sum_{s} p(s) \left(\sum_{i \in s} y_i b_{si} \right)$$

$$= \sum_{i=1}^{N} y_i \left(\sum_{s \ni i} p(s) b_{si} \right)$$

and if $E(t_b)$ is to equal $Y = \sum_{1}^{N} y_i \, \forall \underline{Y} = (y_1, \ldots, y_i, \ldots, y_N)$ then the condition (2.5.1) must hold.

An interesting example of a t_b is the following:

Suppose $\underline{X} = (x_1, \ldots, x_i, \ldots, x_N)$ with $x_i > 0 \, \forall i \in U$ and $X = \sum_{1}^{N} x_i$ be known and

$$(i) p(s) = \frac{\sum_{i \in s} x_i}{X} \cdot \frac{1}{\binom{N-1}{n-1}} \tag{2.5.2}$$

with every s containing $n(1 < n < N)$ distinct units and $b_{si} = \frac{X}{\sum_{i \in s} x_i}$ for every s such that $s \ni i$. Then

$$t_b = X \frac{\sum_{i \in s} y_i}{\sum_{i \in s} x_i} = X \frac{\bar{y}}{\bar{x}},$$

$$\bar{y} = \frac{1}{n} \sum_{i \in s} y_i, \bar{x} = \frac{1}{n} \sum_{i \in s} x_i.$$

That is, t_b is the 'ratio estimator' for Y and it is unbiased for Y if it is based on the design p subject to the condition (2.5.2).

A t_b with the condition (2.5.2) relaxed is called a 'homogeneous linear estimator' (HLE) rather than the 'homogeneous linear unbiased estimator' (HLUE) for Y.

An estimator of the form

$$t_l = a(s) + \sum_{i \in s} y_i b_{si}$$

with $a(s)$ and b_{si} free of $\underline{Y}$ is called a linear estimator and it is linear unbiased estimator (LUE) for Y if in addition b_{si} is subject to (2.5.1) and $a(s)$ is subject to

$$\sum_s a(s)p(s) = 0.$$

2.6 Godambe's and Basu's Theorems on Non-Existence of Unbiased Estimators and Exceptional Situations

Godambe (1955) in fact opened the modern theory of sample surveys by stating and proving the

Theorem 2 *For a sampling design in general among the Homogeneous Linear Unbiased Estimators (HLUE) for a finite population total, there does not exist one with the uniformly minimum variance (UMV).*

Proof For $t_b = \sum\limits_{i \in s} y_i b_{si}$ subject to

$$\sum_s p(s)b_{si} = 1 \,\forall\, i \tag{2.6.1}$$

implying

$$E(t_b) = Y \,\forall\, \underline{Y} = (y_1, \cdots, y_i, \cdots, y_N)$$
$$V(t_b) = E(t_b^2) - Y^2, \text{ let us minimize}$$

$$E(t_b^2) = \sum_s p(s) \left(\sum_{i \in s} y_i b_{si} \right)^2$$

with respect to b_{si} subject to (2.6.1).

For an s containing i with $p(s) > 0$, $y_i \neq 0$, taking λ as a Lagrangian undetermining multiplier let us for this purpose try to solve

$$0 = \frac{\partial}{\partial b_{si}} \left[\sum_s p(s) \left(\sum_{i \in s} y_i b_{si} \right)^2 - \lambda \left(\sum_s p(s)b_{si} - 1 \right) \right]$$

$$= 2y_i p(s) \left(\sum_{i \in s} y_i b_{si} \right) - \lambda p(s)$$

$$\text{or } \sum_{i \in s} y_i b_{si} = \frac{\lambda}{2y_i}.$$

For a particular case of $\underline{Y}$ as

$$\underline{Y}^{(i)} = (0, \ldots, 0, y_i \neq 0, 0, \ldots, 0) \text{ it then follows}$$

$$b_{si} = \frac{\lambda}{2y_i^2} = b_i(\text{say}),$$

Then, (2.6.1) gives

$$b_i = \frac{1}{\pi_i}, \text{ since } \pi_i \neq 0 \text{ by Theorem 1.}$$

So, for the $t(b)$ with a possible UMV, we need

$$t(b) = \sum_{i \in s} \frac{y_i}{\pi_i} = \frac{\lambda}{2y_i} \text{ for } y_i \neq 0, i \in s, \pi_i \neq 0.$$

Then, for an s with $s \ni j, k, \pi_j \neq 0, \pi_k \neq 0$, we must have

$$\frac{y_i}{\pi_i} + \frac{y_j}{\pi_j} + \frac{y_k}{\pi_k} = \frac{\lambda}{2y_i}$$

whatever real numbers y_j, y_k may be. Hence, Godambe concludes that for a general sampling design, this cannot happen and an HLUE with a UMV property cannot exist.

Hege (1965) and Hanurav (1966), however, argue that for an exceptional class of uni-cluster designs (UCD) p with the property that for two samples s_1, s_2 with $p(s_1) > 0, p(s_2) > 0$, either (1) $s_1 \cap s_2 = \Phi$ (the empty set), or (2) $s_1 \sim s_2$ (equivalent), i.e.

$$i \in s_1 \Longleftrightarrow i \in s_2$$

ie, any two positive-probability-valued samples are either disjoint or composed of a same set of distinct units. Such UCD's can easily be constructed as composed of the units like

$$i, K + i, 2K + i, \ldots, (n - 1)K + i, i = 1, 2, \ldots, K,$$

with K and n as positive integers > 1 each.

N. B. So, Godambe's Theorem applies only to non-UCD (NUCD) designs. But whether a UCD design admits a UMV estimator among HLUE's for Y is a question which we cannot answer at this stage, but will be answered in the affirmative later in this book.

After (Godambe, 1955) earth-shaking publication, Basu (1971) came up with another compelling Sample Survey result that except a Census Design, no other design can admit a UMV estimator among all unbiased estimators for a finite population total. For a proof, let us recall the population vector

$$\Omega = \{\underline{Y} = (y_1, \ldots, y_i, \ldots, y_N) \,\big|\, -\infty < a_i \leq y_i \leq b_i < +\infty, \,\forall\, i \in U\}$$

Let $\underline{A} = (A_{1j} \ldots, A_{ij}, \ldots, A_N)$ be any arbitrary point in Ω and $A = \sum_1^N A_i$. Let $t = t(s, \underline{Y})$ be any unbiased estimator for $Y = \sum_1^N y_i$. Let, if possible $t^* = t^*(s, \underline{Y})$ be the UMV estimator for Y so that $E(t^*) = Y = E(t)$ and given any estimator t for Y other than t^* one has

$$V(t^*) \leq V(t)\,\forall\,\underline{Y} \in \Omega$$

and $\qquad\qquad V(t^*) < V(t)$ for at least one element $\underline{Y}$ in Ω.

Let $t_A = t_A(s, \underline{Y})$ be any unbiased estimator for Y defined as

$$t_A(s, \underline{Y}) = t^*(s, \underline{Y}) - t^*(s, \underline{A}) + A$$

Then, $\qquad\qquad\qquad Et_A(s, \underline{Y}) = Y - A + A = Y$

i.e., t_A is unbiased for Y and

$$V(t_A)\big|_{\underline{Y}=\underline{A}} = E\left[t^*(s, \underline{Y}) - t^*(s, \underline{A}) + A - Y\right]^2\big|_{\underline{Y}=\underline{A}}$$
$$= 0.$$

This implies $V(t^*)\big|_{\underline{Y}=\underline{A}} = 0$ too because t^* is the UMV estimator for Y. Similarly, for another point $\underline{B} = (B_1, \ldots, B_i, \ldots, B_N)$ in Ω the estimator $t_B = t_B(s, \underline{Y}) = t^*(s, \underline{Y}) - t^*(s, \underline{B}) + B$ also has $E(t_B) = Y$ and $V(t_B) = \big|_{\underline{Y}=\underline{B}} = 0$ implying $V(t^*) = \big|_{\underline{Y}=\underline{B}} = 0$, i.e., $V(t^*) = 0$ at every point of Ω. This is possible for no sampling design other than the census design p_c, say, which assigns a zero selection probability to every sample other than the sample which coincides with the population U itself and $p_c(U) = 1$ while $p_c(s) = 0\,\forall\,s$ unless $s = U$.

Basu (1971) result does not give us (Godambe, 1955) because starting with an HLUE t_H one cannot establish Basu's result as one cannot construct an estimator like t_A, t_B, etc., as is required to apply (Basu, 1971) proof.

2.7 Sufficiency and Minimal Sufficiency

On $U = (1, \ldots, i, \ldots, N)$, the finite survey population is defined as a real variable y leading to the concept of a parametric space

$$\Omega = \{\underline{Y}\,|\, -\infty < a_i \leq y_i \leq b_i < +\infty, i = 1, \ldots, N\}$$

with a_i, b_i possibly as known real numbers.

For the observable survey data

$$d = (s, y_i | i \in s), s = (i_1, \ldots, i_j, \ldots, i_n)$$

or by more elaborately, $d = \big((i_1, y_{i_1}), \ldots, (i_n, y_{i_n}) \big)$
and $\Omega_d = \big\{ \underline{Y} | (y_{i_1}, \ldots, y_{i_n})$ as in d and
$-\infty < y_K < +\infty, K = 1, \ldots, N$ other than $K = i_1, \ldots, i_n \big\}$

denotes the part of Ω consistent with data d.

Corresponding to s let $s^* = \{ j_i, \ldots, j_K \}$, the set of distinct units in s suppressing the order and frequency of units of U in s and

let $d^* = \big\{ s^*, y_k | k \in s^* \big\} = \big\{ (j_1, y_{j_1}), \cdots, (j_k, y_{j_k}) \big\}$
and $\Omega_{d^*} = \{ \underline{Y} | y_k$ as in d^*, but other y_i's as $-\infty < y_i < +\infty \, \forall i \in s^*, \, y_k \in d^* \}$

i.e., the part of the parametric space consistent with d^*.

However, clearly, $\Omega_d = \Omega_{d^*}$.

Let $I_{\underline{Y}}(d) = 1$ if $\underline{Y} \in \Omega_d$
                                                              $= 0$, else.
Also, let $I_{\underline{Y}}(d^*) = 1$ if $\underline{Y} \in \Omega_{d^*}$
                                                              $= 0$, else.
So, $I_{\underline{Y}}(d) = I_{\underline{Y}}(d^*)$

Let us restrict to only 'non-informative' designs for which $p(s)$ for every s in the sample space, i.e., the collection of all possible s that one may choose, be independent of every element of $\underline{Y}$ in Ω.

Let $P_{\underline{Y}}(d)$ be the probability of observing the data d when $\underline{Y}$ is the underlying parametric point in the parametric space Ω. Then, since we restrict to 'non-informative' design p, we may write

$$P_{\underline{Y}}(d) = p(s) I_{\underline{Y}}(d)$$
Also, $$P_{\underline{Y}}(d^*) = p(s) I_{\underline{Y}}(d^*)$$

Also, $P_{\underline{Y}}(d) = P_{\underline{Y}}(d \cap d^*)$ because d is one of the 'unique' points in the set of one or more points d^* so that $d = d \cap d^*$.

So,
$$P_{\underline{Y}}(d) = P_{\underline{Y}}(d \cap d^*) = P_{\underline{Y}}(d^*)P_{\underline{Y}}(d|d^*)$$
or,
$$p(s)I_{\underline{Y}}(d) = p(s^*)I_{\underline{Y}}(d^*)P_{\underline{Y}}(d|d^*)$$
giving
$$P_{\underline{Y}}(d|d^*) = \frac{p(s)I_{\underline{Y}}(d)}{p(s^*)I_{\underline{Y}}(d^*)}$$

agreeing to suppose $I_{\underline{Y}}(d^*) \neq 0$ because this situation having no relevance of interest. So, $P_{\underline{Y}}(d|d^*) = \frac{p(s)}{p(s^*)}$, which is free of $\underline{Y}$, i.e., a constant.

Hence, given d, the statistic d^* is a 'sufficient statistic'.

Now, let us clarify what we mean by a 'Minimal Sufficient Statistic'. The raw data
$$d = (s, \underline{Y})$$

may be supposed to constitute each a 'point' in the data space $\mathfrak{D} = \{d\}$, the totality of all possible 'data points'. A statistic
$$t = t(d)$$

which is a 'set' of several data points inducing a 'Partition' in the data space. By a 'Partition', we mean a collection of 'data points' which are mutually exclusive and which together are co-extensive with the entire data space. Two different statistics
$$t_1 = t_1(d) \text{ and } t_2 = t_2(d)$$

induce on the data space two different 'Partitioning sets'. If one set of 'Partitioning sets' induced by one statistic t_1 is completely within the one induced by another statistic t_2, then the statistic t_1 is regarded as 'thiner' than t_2 which is 'thicker' than t_1.

Of the two statistics t_1, t_2, the one inducing a thicker partitioning on the data space archives a greater summarization. If they are again both sufficient statistics then between them the one inducing a thicker partitioning on $\mathfrak{D}$, the data space is to be preferred because neither of them sacrifices any relevant information, and the one inducing the thicker partitioning achieves a higher summarization and sacrifices no information. So, it is highly desirable to find the 'Minimal Sufficient Statistic', if it exists, because it sacrifices no information and induces the 'thickest partitioning' or achieves the highest level of summarization of data among all those sufficient statistics none of which sacrifices any relevant information. So, it is of interest to find 'Minimal Sufficient' statistic in the context of sample surveys.

Let d_1 and d_2 be two elementary data points and d_1^* and d_2^* be the corresponding sufficient statistics as d^* corresponds to d. Let t be another statistic other than d^* if such a one exists. Let $t(d_1)$ and $t(d_2)$ be sufficient statistics corresponding to d_1 and d_2.

Let, in particular,
$$t(d_1) = t(d_2).$$

Thus, $t(d_1)$ and $t(d_2)$ are points in the same 'Partitioning Set' induced on $\mathfrak{D}$ by t. If we can show that d_1^* and d_2^* also are in the same partitioning set induced on $\mathfrak{D}$ by d^*, then it will follow that the partitioning set induced by t is 'within' the partitioning set induced by d^* implying that d^* induces on $\mathfrak{D}$ a thicker partition set than the one induced on it by t. Since both t and d^* are two different sufficient statistics and t is any sufficient statistic other than d^* it will follow that d^* is the minimal sufficient statistic inducing on $\mathfrak{D}$, the thickest partitioning sacrificing no relevant information and achieving the maximal level of summarization of data. Let us prove this formally and analytically.

By definition

$$
\begin{aligned}
P_{\underline{Y}}(d_1) &= P_{\underline{Y}}(d_1 \cap t(d_1)) \\
&= P_{\underline{Y}}(t(d_1)) P_{\underline{Y}}(d_1 | t(d_1)) \\
&= P_{\underline{Y}}(t(d_1)) C_1
\end{aligned}
$$

with C_1 as a constant free of $\underline{Y}$ because t is a sufficient statistic.

Also, $P_{\underline{Y}}(d_2) = P_{\underline{Y}}(t(d_2)) C_2$ with C_2 as another constant.

Since, $t(d_1) = t(d_2)$, it follows that

$$
P_{\underline{Y}}(d_1) = P_{\underline{Y}}(t(d_1)) C_1 = P_{\underline{Y}}(t(d_2)) C_1 = P_{\underline{Y}}(d_2) \frac{C_1}{C_2}
$$

or,
$$
p(s_1) I_{\underline{Y}}(d_1) = p(s_2) I_{\underline{Y}}(d_2) \frac{C_1}{C_2}
$$

Also,
$$
p(s_1^*) I_{\underline{Y}}(d_1^*) = p(s_1) I_{\underline{Y}}(d_1) C_3
$$

with C_3 also a constant because d_1^* is a sufficient statistic.

Similarly, $p(s_2^*) I_{\underline{Y}}(d_2^*) = p(s_2) I_{\underline{Y}}(d_2) C_4$ with C_4 another constant.

So,
$$
p(s_1^*) I_{\underline{Y}}(d_1^*) = p(s_2^*) I_{\underline{Y}}(d_2^*) \frac{C_1 C_3}{C_2 C_4}
$$

So,
$$
I_{\underline{Y}}(d_1^*) = I_{\underline{Y}}(d_2*)
$$

and hence $d_1^* = d_2^*$.

So d^* is the 'Minimal Sufficient statistic' corresponding to d.

2.8   Complete Class Theorem

Let us suppose that starting with any initial sample s with $p(s)$ as its non-informative selection-probabilities, it is possible to derive a sample from s as s^*, which is free of the order and/or multiplicity of the units in s. Then, given an estimator $t = t(s, \underline{Y})$, let us construct the estimator

$$t^* = t^*(s, \underline{Y}) = t^*(s^*, \underline{Y})$$

for every s to which the same s^* corresponds and is given by

$$t^* = t^*(s^*, \underline{Y}) = \frac{\sum\limits_{s \to s^*} p(s)t(s, \underline{Y})}{\sum\limits_{s \to s^*} p(s)} = t^*(s, \underline{Y})$$

for every s to which corresponds the same s^*. Here $\sum_{s \to s^*}$ denotes sum over those samples s to each of which the same s^* corresponds.

Then,
$$E(t^*) = \sum_s p(s)t^*(s, \underline{Y}$$

$$= \sum_{s^*} \left[\sum_{s \to s^*} p(s)t^*(s^*, \underline{Y}) \right]$$

$$= \sum_{s^*} \left[\sum_{s \to s^*} p(s) \frac{\sum\limits_{s \to s^*} p(s)t(s, \underline{Y})}{\sum\limits_{s \to s^*} p(s)} \right]$$

$$= \sum_{s^*} \left[\sum_{s \to s^*} p(s)t(s, \underline{Y}) \right]$$

$$= \sum_s p(s)t(s, \underline{Y})$$

$$= E_p(t)$$

Next,
$$V(t) = E[t - E(t)]^2$$
$$V(t^*) = E[t^* - E(t^*)]^2$$
So,
$$V(t) = E[t - E(t^*)]^2$$

Now,
$$E(tt^*) = \sum_s p(s)t^*(s^*, \underline{Y})t(s, Y)$$

$$= \sum_{s^*} t^*(s^*, \underline{Y}) \left[\frac{\sum\limits_{s \to s^*} p(s)t(s, \underline{Y})}{\sum\limits_{s \to s^*} p(s)} \right] \sum_{s \to s^*} p(s)$$

$$= \sum_{s^*} \left[t^*(s^*, \underline{Y}) \right]^2 p(s^*) = E\left[(t^*)^2 \right]$$

writing $p(s^*) = \sum\limits_{s \to s^*} p(s)$.

So,
$$E(t - t^*)^2 = E(t^2) + E(t^*)^2 - 2E(tt^*)$$
$$= E(t^2) - E(t^*)^2$$
$$= V(t) - V(t^*) \text{ because } E(t) = E(t^*)$$

So, $V(t^*) = V(t) - E(t - t^*)^2 \le V(t)$

This inequality is strict unless $t = t^*$ with probability one, i.e., t and t^* are of the same form.

So, given an unbiased estimator $t = t(s, \underline{Y})$ for a finite population total if s^* exists different from s an estimator based on s^* ie, $t = t(s^*, \underline{Y})$ can be constructed with a uniformly smaller variance.

Because of this result, emerges the following interesting concept.

Within a wide class of estimators, say, among all unbiased estimators of a finite population total, a particular class of estimators may be constructed such that given any estimator outside the latter 'sub-class' but within the former wider class, one estimator 'better than' this one in terms of variance, may be found to exist. This sub-class is composed of the better estimators than the ones outside, therefore, it is called a 'Complete Class'. Thus, given an 'unbiased estimator', which is not a function of the 'minimal sufficient statistic', one with a uniformly smaller variance may be found as one which is a function of the 'minimal sufficient statistic' as clearly illustrated using the above procedure of construction, say, if t^* versus the original t as an initial estimator.

This 'Complete Class Theorem' is very useful in establishing, as shown just below, that a UCD admits a unique UMV estimator among all 'Homogeneous Linear Unbiased Estimators' (HLUE) for a finite population total.

Let $t_b = \sum\limits_{i \in s} y_i b_{s_i}$ be an HLUE for $Y = \sum\limits_{1}^{N} y_i$ subject to $\sum\limits_{s \ni i} p(s) b_{s_i} = 1 \forall i \in U$ with b_{s_i}'s free of $\underline{Y} = (y_1, \ldots, y_i, \ldots, y_N)$ based on a sample s chosen with probability $p(s)$, when the design p is a UCD satisfying the condition s_1, s_2 with $p(s_1) > 0$, $p(s_2) > 0$ for which 'either' $s_1 \cap s_2 = \Phi$ (the empty set) 'or' $s_1 \sim s_2$, i.e., $i \in s_1 \iff i \in s_2$, i.e., all the units in s_1 are common with those in s_2 and vice versa.

Let further, t be a function of the 'minimal sufficient statistic' d^*. Then, if there is to exist the UMV estimator (UMVE) t^* for Y, then it must be a member of the 'Complete Class' of statistics, i.e., t_b must be a function of d^*. Since t_b is in the Complete Class and the design is UCD, for every s containing a unit, i must contain no other unit and b_{s_i} must be of the form b_i because else t_b cannot be in the 'Complete Class' and so we must have

$$1 = b_i \left(\sum\limits_{s \ni i} p(s) \right) = b_i \pi_i$$

or $b_i = \frac{1}{\pi_i}$ (this $\pi_i > 0$ because otherwise t_b cannot be unbiased for Y) and so $t_b = \sum_{i \in s} \frac{y_i}{\pi_i}$, i.e., the 'Complete Class of unbiased estimator of Y based on a UCD' must consist of the unique estimator which is the 'Horvitz–Thompson' estimator, and hence this $t_H = \sum_{i \in s} \frac{y_i}{\pi_i}$ must be the unique UMVUE of Y based on a UCD design.

N. B. Through these Godambe's result has an exception for a UCD; this unique UMVUE result is too insignificant because this UMVUE has no competitor worthy of attention.

2.9 Comparison Between SRSWR and SRSWOR

If from the population $U = (1, \ldots, i, \ldots, N)$ of N units, an SRSWR is taken in n draws and also an SRSWOR is taken in n draws, then we have already seen that the corresponding sample means $\bar{y}_{WR} = \frac{1}{n} \sum_{k=1}^{n} y_k$, here y_k denoting the value of y among $\underline{Y} = (y_1, \ldots, y_i, \ldots, y_N)$ for the units chosen on the Kth draw $(K = 1, \ldots, n)$ and, respectively, $\bar{y}_{WOR} = \frac{1}{n} \sum_{i \in s} y_i$, then we have seen that

$$V(\bar{y}_{WR}) = \frac{\sigma^2}{n}, \qquad \sigma^2 = \frac{1}{N} \sum_{1}^{N} (y_i - \underline{Y})^2 \text{ and}$$

$$V(\bar{y}_{WOR}) = \frac{N - n}{N} \frac{S^2}{n}, \qquad S^2 = \frac{1}{N - 1} \sum_{1}^{N} (y_i - \underline{Y})^2.$$

Since $\sigma^2 = \frac{N-1}{N} S^2$, clearly,

$$V(\bar{y}_{WR}) = \frac{N - 1}{N} \frac{S^2}{n} > V(\bar{y}_{WOR}) \; \forall n \geq 2.$$

Thus, SRSWOR is better than SRSWR.

But this comparison is unfair because in SRSWR the number of distinct units which only need to be surveyed may be smaller in varying degrees for various values of n as we have seen that the expected number of distinct units in an SRSWR in n draws, i.e., (N, n) is

$$v = N \left[1 - \left(\frac{N - 1}{N} \right)^n \right]$$

$$= N \left[1 - \left(1 - \frac{1}{N} \right)^n \right] = N \left[\frac{n}{N} - \frac{n(n - 1)}{2N^2} + \cdots \right] < n.$$

Since our book, cf, Chaudhuri (2010) may be treated as a 'companion book' vis-a-vis this one, it is straightaway recommended that the pp. 32–41 in Chaudhuri (2010) should be consulted carefully to make a rational study of a comprehensive comparison 'SRSWR vs SRSWOR'.

But in the present book, we may point out that $\bar{y}_{\text{WOR}} = \frac{1}{N} \sum_{i \in s} \frac{y_i}{\pi_i(\text{WOR})}$ because

$$\pi_i(\text{WOR}) = \frac{n}{N} \text{ while } \bar{y}_{\text{WR}} = \frac{1}{n} \sum_{k=1}^{n} y_k \text{ but}$$

$$\frac{1}{N} t_{\text{WR}}(HT) = \frac{1}{N} \sum_{i \in s} \frac{y_i}{\pi_i(\text{WR})}$$

is an unbiased estimator for $\bar{Y}$ and

$$\pi_i(\text{WR}) = 1 - \left(\frac{N-1}{N}\right)^n \quad \forall i \in U.$$

So, $\bar{y}_{\text{WR}}$ versus $\bar{y}_{\text{WOR}}$ one may examine the performances of $\frac{1}{N} t_{\text{WR}}(HT)$ based on SRSWR in order to take another rational view about the performances of SRSWR versus SRSWOR. This is especially because the 2nd order inclusion probabilities

$$\pi_{ij} = 1 - 2\left(\frac{N-1}{N}\right)^n + \left(\frac{N-2}{N}\right)^n$$

for SRSWR are provided so as to facilitate the computation of $V\left(\frac{1}{N} t_{WR}(HT)\right)$ with little difficulty so as to compare it's magnitudes vis-a-vis its competitor $\bar{y}_{\text{WOR}}$ based on SRSWOR.

2.10 Roles of Labels and of Distinct Units in 'With Replacement' Sample

Given the detailed survey data $d = \left((i_1, y_{i_1}), \ldots, (i_j, y_{i_j}), \ldots, (i_n, y_{i_n})\right)$ based on a sample $s = (i_1, \ldots, i_j, \ldots, i_n)$ chosen with a probability $p(s)$ according to a non-informative sampling design p, we have seen that

(1) the function of d, namely $d^* = (s^*, y_i | i \in s^*)$ when $s^* = \{j_1, \ldots, j_K\}$ is the set of $K(1 \leq K \leq n)$ distinct units in s ignoring the order in which the distinct elements $j_1, \ldots, j_K, (1 \leq K \leq n)$ occur in s and s^*, constitutes the minimal sufficient statistic corresponding to d and

(2)

$$t^* = t^*(s, \underline{Y}) = \frac{\sum_{s \longrightarrow s^*} p(s) t(s, \underline{Y})}{\sum_{s \longrightarrow s^*} p(s)} = t^*(s^*, \underline{Y})$$

for every s corresponding to s^* satisfies

$$(i) E(t) = E(t^*)$$

and $$(ii) V(t) - V(t^*) = E(t - t^*)^2$$

and hence $$V(t) \geq V(t^*)$$

and strictly so unless

$$\text{Prob}(t = t^*) = 1$$

for every t corresponding to t^*.

Now since the labels in s^* are inalienable in minimal sufficient statistic $d^* = (s^*, y_i | i \in s^*)$, the labels cannot be ignored in developing a 'Design-based' theory of survey sampling because omitting them from d^* we cannot retain a sufficient statistic without sacrificing efficiency in estimation, essentially recognizing the above as a 'Survey Sampling' version of the celebrated Rao-Blackwell theory in generalization in statistical theory.

But a hindrance in appropriate 'Likelihood-based inference' is easily understandable as we may see below.

We have seen that the likelihood based on d and on d^* takes the forms

$$L(\underline{Y}|d) = p(s) I_{\underline{Y}}(d) \quad \text{and}$$
$$L(\underline{Y}|d^*) = p(s^*) I_{\underline{Y}}(d^*)$$

with $I_{\underline{Y}}(d) = I_{\underline{Y}}(d^*)$ for a non-informative design p. Thus, these likelihoods are 'Flat' and there is no discrimination in the likelihood with respect to the 'sample-unobserved' elements of $\underline{Y}$ in $\Omega = \{\underline{Y} | -\infty < a_i \leq y_i \leq b_i < +\infty\}$ for $i \in U$.

So, using the entire available data d or d^* involving the sampled 'labels', it is impossible to find a 'Maximum Likelihood Estimator', say, for any parametric function $\tau(\theta)$ for $\theta = \theta(\underline{Y})$ with $\underline{Y} \in \Omega$.

So, for a non-informative design, when the entire minimal sufficient statistic d^* is at hand, we cannot ignore labels that are thus informative in inference-making, but on the other hand, they render 'likelihood-based' inferencing impossible as the detailed data based on sufficient statistic-based likelihood is flat and hence sterile.

Hartley et al. (1969), however, derived a Maximum Likelihood Estimator (MLE) for a finite population total $Y = \sum_1^N y_i$ from an SRSWOR, (N, n), with the following formulation of their own. They supposed that the elements of $\underline{Y}$ in Ω cannot be so wide as any arbitrary real numbers but each y_i in $\underline{Y}$ must be just one of a finite number of specified real numbers $z_1, \ldots, z_K$, with K as quite a large number, may be exceeding N so that in $\underline{Y}$, every z_j occurs with an unknown frequency, they call 'Loading' which is N_j such that $\sum_{j=1}^K N_j = N$ and consequently

$Y = \sum_{1}^{N} y_i = \sum_{j=1}^{K} N_j z_j$. With this 'Scale-load' approach with the y_i's scaled as the discrete set of numbers $z_1, \ldots, z_j, \ldots, z_K$ with the respective unknown loadings $N_j (j = 1, \ldots, K)$, given the survey data from an SRSWOR, (N, n), with the observed scale-values z_j (for $j \in s$) with the respective loadings n_j, say, for $j \in s$ so that $n = \sum_{j=1}^{K} n_j$ as observed. Hartley et al. (1969) considered the likelihood for $\underline{N} = (N_1, \ldots, N_j, \ldots, N_K)$ based on an SRSWOR, (N, n) as

$$L(\underline{N}) = \Pi_{j=1}^{K} \binom{N_j}{n_j} \Big/ \binom{N}{n}.$$

$$\text{So,} \quad \widehat{Y}_L = \sum_{j \in s} \widehat{N}_j z_j \text{ with } \widehat{N}_j = \frac{n_j}{n} N$$

is the MLE for $Y = \sum_{1}^{N} y_i = \sum_{j=1}^{K} N_j z_j$.

This theory of Hartley and JNK Rao was not cordially accepted initially by the then experts like (Godambe, 1970, 1975). But with advancing time, it is being accepted. Importantly (Owen, 2001), however, acclaims it and encourages developing inference theories through this discrete scale-load approach, especially in his currently flourishing Empirical Likelihood approach.

In SRSWR, (N, n), the fact that all the units selected in the sample $s = (i_1, \ldots, i_j, \ldots, i_n)$ are not necessarily distinct provides a source for improving on inference-making. From C. R. Rao's authentication, it is well-known that an NSSO worker late Raja Rao out of his own curiosity empirically observed that the mean of all the observed y-values in an SRSWR, (N, n) turns out to have a larger variance than the mean of all the y-values in the set of distinct units found in the SRSWR, (N, n). This gentleman approached Professor C.R. Rao for a possible explanation. But C.R. Rao directed him to Prof. D. Basu for the latter's help in the matter. D. Basu immediately realized that the property of sufficiency in the y-values of the distinct sample-units vis-a-vis the entire sequence of y-values in the selected sample was behind this phenomenon. In the companion book 'Essentials of Survey Sampling' by Chaudhuri (2010) that should be read side by side the present treatise the materials on pp. 33–41 interesting relevant details in which should be drawn to a reader's attention. Further materials on pp. 20–24 in Chaudhuri (2014) text 'Modern Survey Sampling' are also worthy of attention in this context. Repetition is avoided to save space. Chaudhuri and Mukerjee (1984) narrated unbiased estimation of the unknown number of units of a specified category under the SRSWOR scheme.

2.11 Controlled Sampling

Goodman and Kish (1950) introduced a technique of 'Controlled Sampling' to modify SRSWOR for a certain reason. A technical term 'Support' means the set of
samples each with a positive probability of getting selected. 'Support Size' means
the 'number of' samples each carrying a positive selection probability for a particular sampling design. For SRSWOR, with N and n, respectively, as the sizes of the
population and a sample, this number is of course $\binom{N}{n}$ and every sample s has the
common probability of selection, namely $\dfrac{1}{\binom{N}{n}}$. At least in situations when neither
N nor n is small or at least not moderately so, often one may encounter in practice
certain samples which are inconvenient to survey because of inaccessible location,
say, in deep forests, up the hills, in marshy lands, in areas infested with hooligans
or notorious elements of the societies and too far from the convenient headquarters
from where the survey operation may be controlled. In such situations, (Goodman
and Kish, 1950) suggested that it may be scrupulous to avoid selection activities
with respect to such 'Non-preferred Samples', assigned together collectively a small
aggregate selection probability α, say, in magnitude $0 < \alpha \leq 0.05$. But while so
doing, precaution must be taken that for the revised sampling design q, say, $q(s)$
should be positive for a number of samples less than the support size of the initially
contemplated SRSWOR design p, say. Keeping the inclusion probabilities for every
pair (i, j) and every single unit i, the same in both the original and the revised designs
p and q. This restriction is desirably imposed in order to retain the properties of the
original estimator in tact for the designs. This, if the starting design is SRSWOR
employing the sample mean

$$\bar{y} = \frac{1}{n} \sum_{i \in s} y_i$$

which is unbiased for $\bar{Y} = \frac{1}{N} \sum_{1}^{N} y_i$ and has the variance

$$V(\bar{y}) = \frac{(N-n)}{Nn} \frac{1}{N-1} \sum_{1}^{N} (y_i - \bar{Y})^2$$

and the inclusion probabilities $\pi_i = \frac{n}{N} \, \forall i$ in $U = (1, \cdots, i, \cdots, N)$ and
$\pi_{ij} = \frac{n(n-1)}{N(N-1)} \, \forall i, j (i \neq j)$ in U.

If for the revised design q, these π_i and π_{ij}-values are retained, then for the
sample mean $\bar{y}$ based on q also one retains $E(\bar{y}) = \bar{Y}$ and $V(\bar{y}) = \frac{(N-n)}{Nn} S^2$, writing
$S^2 = \frac{1}{N-1} \sum_{1}^{N} (y_i - \bar{Y})^2$. Since π_{ij}'s for q, i.e., $\sum_{s \ni j} q(s) = \pi_j$ are to be $\frac{n(n-1)}{N(N-1)} \, \forall i \neq j$

in U, the number of constraints to which q is subject is $\binom{N}{2}$. Then the 'Support Size'

of q cannot be more than $\binom{N}{n} - \binom{N}{2}$. Thus, the support size $\binom{N}{n}$ for SRSWOR is

reduced for the design q. If one may identify the set of 'Non-preferred Samples', then
one may assign zero selecting probability to a number of samples in the 'Support'
of SRSWOR, (N, n), then keep 'sum of the selection probabilities' assignable to
the 'Non-preferred Samples' the value $\alpha(0 < \alpha \leq 0.05)$ and assign the selection
probabilities suitably to the complementary set of preferred samples ensuring $\pi_i = \frac{n}{N} \forall i \in U$ and $\pi_{ij} = \frac{n(n-1)}{N(N-1)}$, then we may say that we have hit upon a Controlled
Sampling scheme revising the SRSWOR, (N, n) retaining the 'unbiasedness' of $\bar{y}$
and its efficiency in terms of the variance intact vis-a-vis the SRSWOR, (N, n).

To accomplish this algebraically is not easy. The eminent expert in Design of
Experiments, namely late Professor Chakrabarti (1963) gave a simple algorithm
based on the theory of Balanced Incomplete Block Design (BIBD) as follows. In a
BIBD, v treatments are laid out in b blocks each composed of K units, each treatment
occurring r times and every distinct pair of treatments occurring a common number
λ of times so that $bK = vr$ and $\lambda(v - 1) = r(K - 1)$. To see the link of a BIBD
with SRSWOR, let $v = N$, $K = n$ so that it follows that

$$r = b\frac{n}{N} = \frac{\lambda(N - 1)}{(n - 1)} \text{ or } \lambda = \frac{n(n - 1)}{N(N - 1)}b.$$

So, if such a BIBD is constructed following Chakrabarti, b may be treated as the
Support Size for a controlled sampling design vis-a-vis SRSWOR, (N, n) and one
may choose randomly one of the b blocks to get a controlled SRSWOR achieving

$$\pi_{ij} = \frac{n(n - 1)}{N(N - 1)} = \frac{\lambda}{b} \forall i, j (i \neq j = 1, \ldots, N)$$
$$\text{and} \quad \pi_i = \frac{n}{N} = \frac{r}{b} \forall i (= 1, \ldots, N)$$

Here, of course, $b < \binom{N}{n}$, as the Support Size of a Controlled Sampling scheme

derived from SRSWOR which is $\binom{N}{n}$.

But an insurmountable problem appears because for many values of b, a BIBD

may not exist, for example, for $b < \binom{8}{3} = 56$ as is known from the literature in
BIBD.

So, 'Controlled Sampling' as a topic of applied research in sampling is receiv-
ing attention among vigorous researchers. Chaudhuri (2014, 2018), Chaudhuri and
Vos (1988) and Gupta et al. (2012) are important sources for further information
on this topic. Chaudhuri (2014, 2018) has mentioned that (Rao and Nigam, 1990,

1992) provided a decisive solution based on Simplex method applicable in Linear Programming problems. One cannot go into these at this stage and further coverage needs to be postponed to a later part of the present volume.

2.12 Minimax Approach

For an estimator $t = t(s, \underline{Y})$ for $Y = \sum_1^N y_i$ based on s with selection probability $p(s)$ with the Mean Square Error (MSE)

$$M(t) = E(t - Y)^2 = \sum_s p(s)(t - Y)^2$$

$$= \sum_s p(s)(t(s, \underline{Y}) - Y)^2$$

$$= \sum_s p(s)A(s), \text{ say, writing}$$

$$A(s) = (t(s, \underline{Y}) - Y)^2$$

the sampling design p will be said to provide the 'Minimax' solution if the maximum possible value of the above $M(t)$ for a fixed choice of t is made the minimum for a specific choice of p.

Chaudhuri (1969) came out with the following inequality: Given the real numbers $(a_i, b_i), i = 1, \ldots, T$ satisfying

(i) $a_1 \geq a_2 \geq \cdots, \geq a_T \geq 0$ and
(ii) $b_1 \geq b_2 \geq \cdots, \geq b_T \geq 0$,
     then, for any permutation $(\partial_1, \ldots, \partial_T)$ of $1, \ldots, T$ it follows that
(iii) $\sum_1^T a_i b_i \geq \frac{1}{T} \left(\sum_1^T a_i \right) \left(\sum_1^T b_i \right)$ and
(iv) $\sum_1^T a_i b_{T-i+1} \leq \sum_1^T a_j b_{ji} \leq \sum_1^T a_i b_i$.

The inequality (iii) clearly implies that the design p for which, given N and n, $p(s) = \dfrac{1}{\binom{N}{n}}$ for every s of a size n provides a 'Minimax' solution because $\sum_s p(s)a(s)$ becomes the maximum if $a(s)$ and $p(s)$ are 'similarly ordered' and that 'highest magnitude' can be controlled to its smallest value if every $p(s)$ is taken as $p(s) = \dfrac{1}{\binom{N}{n}}$ for each of the $\binom{N}{n}$ samples of size n from the population of size N.

2.13 Lessons and Exercises

Lessons with solutions

1. Consider a population $U = (1, \ldots, i, \ldots, N)$ of N units composed of the K disjoint(displayed below)clusters C_i of $n = \frac{N}{K}$ units in K columns $C_1, \ldots, C_2, \ldots, C_i, \ldots, C_K$.

$$
\begin{pmatrix} 1 \\ K+1 \\ \vdots \\ (n-1)K+1 \end{pmatrix} \begin{pmatrix} 2 \\ K+2 \\ \vdots \\ (n-1)K+2 \end{pmatrix} \cdots \begin{pmatrix} i \\ K+i \\ \vdots \\ (n-1)K+i \end{pmatrix} \cdots \begin{pmatrix} K \\ 2K \\ \vdots \\ nK \end{pmatrix}
$$
$$
\;\;\;\;C_1 \qquad\qquad C_2 \qquad\qquad \cdots C_i \qquad\qquad \cdots C_K \;\;.
$$

Consider the design p for which $\mathrm{Prob}(C_i) = p(C_i) = \frac{1}{K}, i = 1, \ldots, K$. Consider estimating the population mean $\bar{Y} = \frac{1}{N}\sum_{j=1}^{N} y_j = \frac{1}{nK}\sum_{j=0}^{n-1}\sum_{i=1}^{K}(y_{jK+i})$ by the sample mean

$$
\bar{y}_i = \frac{1}{n}\sum_{j=0}^{n-1} y_{jK+i}
$$

based on the sample which is the ith cluster if selected with probability $p(C_i) = \frac{1}{K}$ for any of the K clusters $C_i, i = 1, \ldots, K$.

Show that (i) this $\bar{y}_i$ is unbiased for $\bar{Y}$ but the (ii) variance of it cannot be unbiasedly estimated.

$$
E(\bar{y}_i) = \frac{1}{K}\sum_{i=1}^{K} \bar{y}_i = \frac{1}{K}\sum_{i=1}^{K}\left(\frac{1}{n}\sum_{j=0}^{n-1} y_{jK+i}\right)
$$

$$
= \frac{1}{nK}\sum_{i=1}^{K}\sum_{j=0}^{n-1} y_{jK+i} = \bar{Y} \,\forall i = 1, \ldots, K.
$$

Thus, $\bar{y}_i$ based on y-values for the units in a sampled cluster C_i is unbiased for $\bar{Y}$.

Now $V(\bar{y}_i) = E(\bar{y}_i)^2 - \bar{Y}^2$

$$= E(\bar{y}_i)^2 - \frac{1}{N^2}\left[\sum_{i=1}^{K}\sum_{j=0}^{n-1} y_{jK+i}\right]^2$$

$$= E(\bar{y}_i)^2 - \frac{1}{N^2}\left[\sum_{r=1}^{N} y_r\right]^2$$

$$= E(\bar{y}_i)^2 - \frac{1}{N^2}\left[\sum_{r=1}^{N} y_r^2 + \sum_{r\neq r'}^{N}\sum^{N} y_r y_r'\right]$$

$$= E(\bar{y}_i)^2 - \frac{1}{N^2}\left[\sum_{i=1}^{K}\sum_{j=0}^{n-1} y_{jK+i}^2 + \sum_{\substack{i=1 \\ i\neq i'}}^{K}\sum_{i'=1}^{K}\sum_{\substack{j=0 \\ j\neq j'}}^{n-1}\sum_{j'=0}^{n-1} y_{jK+i}y_{j'K'+i'}\cdot\right]$$

The term $\sum_{i=1}^{K}\sum_{j=0}^{n-1} y_{jK+i}^2$ has an unbiased estimator $\frac{1}{K}\sum_{j=0}^{n-1} y_{jK+i}^2$. But the 2nd term within the parentheses cannot be unbiasedly estimated because every term involving y_r's for units in two different clusters have a zero inclusion-probability in any single cluster C_i sampled. So $V(\bar{y}_i)$ cannot be estimated by $(\bar{y}_i)^2$—an unbiased estimator for $(\bar{Y})^2$.

Exercises with hints

1. Is Godambe's theorem a corollary to Basu's?
   Basu (1971) theorem says that among all unbiased estimators for a finite population, total none exists with a uniformly least variance for any non-census design. But Godambe's theorem refers only to the sub-class of homogeneous linear unbiased estimators. Yet this does not follow from Basu's theorem because to prove Basu's theorem, one needs to use a non-homogeneous linear unbiased estimator. This is not permitted if one confines to Godambe (1955) homogeneous linear unbiased estimators.

2. Given

   (i) $a_1 \geq \cdots, \geq a_N \geq 0$ and
   (ii) $b_1 \geq \cdots, \geq b_N \geq 0$ prove that

   (A) $\sum_{i=1}^{N} a_i b_i \geq \frac{1}{N}\left(\sum_{i=1}^{N} a_i\right)\left(\sum_{i=1}^{N} b_i\right)$ and

   (B) $\sum_{i=1}^{N} a_i b_{N-i+1} \leq \sum_{i=1}^{N} a_i b_{ji} \leq sum_{i=1}^{N} a_i b_i$ if $(j_1, \ldots, j_N)$ is a permutation of $(1, \ldots, N)$.

   Solution: See Chaudhuri (1969).

With the background so far developed, no feasible project work could be undertaken to develop any practical work to report.

References

Basu, D. (1971). An essay on the logical foundation of statistical inference. In V. P. Godambe & D. A. Sprout (Eds.) (pp. 203–242). Toronto, Canada: Holt, Rinehart & Winston.

Chakrabarti, M. C. (1963). On the use of incidence matrices in sampling from finite populations. *JISA, 1*, 78–85.

Chaudhuri, A. (1969). Minimax solutions of some problems in sampling from finite populations. *CSA Bulletin, 18*, 1–24.

Chaudhuri, A., & Mukerjee, R. (1984). Unbiased estimation of domain parameters in sampling without replacement. *Survey Methodology, 10*(12), 181–185.

Chaudhuri, A. (2010). *Essentials of survey sampling*. New Delhi, India: PHI.

Chaudhuri, A. (2014). *Modern survey sampling*. Florida, USA: CRC Press.

Chaudhuri, A. (2018). *Survey Sampling*. Florida, USA: CRC Press.

Chaudhuri, A., & Vos, J. W. E. (1988). *Unified theory and strategies of survey sampling*. North Holland, Amsterdam, The Netherlands: Elsevier Science.

Godambe, V. P. (1955). A unified theory of sampling from finite populations. *JRSS B, 17*, 269–278.

Godambe, V. P. (1970). Foundations of survey-sampling. *The American Statistician, 24*, 33–38.

Godambe, V. P. (1975). A reply to my critics. *Sankhyā, C, 37*, 53–76.

Goodman, L. A., & Kish, L. (1950). Controlled selection a technique in probability sampling. *JASA, 55*, 350–372.

Gupta, V. K., Mandal, B. N., & Parsad, R. (2012). *Combinatorics in sample surveys vis-a-vis controlled selection*. Searbrucken, Germany: Lambert Academic Publ.

Hanurav, T. V. (1966). Some aspects of unified sampling theory. *Sankhya A, 28*, 175–204.

Hartley, H. O., & Rao, J. N. K. (1969). A new estimation theory for sample surveys, II. *New developments in survey sampling* (pp. 147–169). New York: Wiley Interscience.

Hege, V. S. (1965). Sampling designs which admit uniformly minimum variance unbiased estimators. *CSA Bulletin, 14*, 160–162.

Owen, A. B. (2001). *Empirical likelihood*. Florida, NY, USA: CRC Press.

Rao, J. N. K., & Nigam, A. K. (1990). Optimal controlled sampling designs. *Biometrika, 77*, 807–814.

Rao, J. N. K., & Nigam, A. K. (1992). Optimal controlled sampling: A unified approach. *International Statistical Review, 60*, 89–98.

Chapter 3
Sampling with Varying Probabilities

3.1 Stratified Sampling

The first prominent unequal probability sampling is by 'stratification'. We have seen Simple Random Sampling (SRS) is useful in making inferences about a population. To estimate the 'parameter', namely the population total or the mean, the SRS is useful in providing an unbiased estimator, namely the sample mean $\bar{y}$ for the population mean $\bar{Y}$ and likewise the expansion estimator $N\bar{y}$ for the population total $Y = N\bar{Y}$. Also, separately for SRSWR and SRSWOR, we have seen their measures of error, namely $V(\bar{y})$ as well as their unbiased estimators in the respective cases of WR and WOR in SRS. But, importantly, we have observed that the variances decrease with the rise in the sample-size, other population-related things remaining intact. So, except exercising a freedom in controlling the sample-size, we have no way to have a conscious effort to regulate or control the efficacy in our estimation procedure by dint of SRSWR or SPSWOR. Stratification is an early attempt to control or regularize efficiency in estimation by a conscious way in changing sample selection as well as estimation method in sampling strategies. Let us discuss this at some length.

Suppose a finite population of N units is split up into a number of H of disjoint parts, respectively, composed of N_h units, $h = 1, \ldots, H, \sum_{h=1}^{H} N_h = N$. From each of these H parts, let SRSWORs of n_h units be, respectively, selected independently of each other with each n_h at least equal to one. Then these parts are called strata, H in number and such a kind of sampling is called Stratified Simple Random Sampling Without Replacement.

Let $\bar{y}_h$ be the sample mean of the n_h units drawn from the hth stratum, $h = 1, \ldots, H$ as, say,

$$\bar{y}_h = \frac{1}{n_h} \sum_{i \in s_h} y_{h_i}$$

writing s_h for the SRSWOR of n_h units drawn from the hth stratum and y_{h_i} for the value of y-variable of interest for the ith unit of the hth stratum.

© The Author(s), under exclusive license to Springer Nature Singapore Pte Ltd. 2022
A. Chaudhuri and S. Pal, *A Comprehensive Textbook on Sample Surveys*, Indian Statistical Institute Series, https://doi.org/10.1007/978-981-19-1418-8_3

Then, obviously

$$E(\bar{y}_h) = \frac{1}{N_h} \sum_{i=1}^{N_h} y_{h_i} = \bar{Y}_h, \text{ say,}$$

which is the mean of all the N_h values of y in the hth stratum.

Also,

$$V(\bar{y}_h) = \frac{N_h - n_h}{N_h n_h} \sum_{i=1}^{N_h} \frac{\left(y_{h_i} - \bar{Y}_h\right)^2}{N_h - 1}$$

$$= \left(\frac{1}{n_h} - \frac{1}{N_h}\right) S_h^2, \text{ writing}$$

$$S_h^2 = \frac{1}{N_h - 1} \sum_{i=1}^{N_h} \left(y_{h_i} - \bar{Y}_h\right)^2,$$

the variance of all the N_h values of y in the hth stratum.

Further,

$$\bar{y}_{st} = \frac{1}{N} \sum_{h=1}^{H} N_h \bar{y}_h$$

has

$$E(\bar{y}_{st}) = \frac{1}{N} \sum_{h=1}^{H} N_h \bar{Y}_h$$

$$= \frac{1}{N} \sum_{h=1}^{H} \sum_{i=1}^{N_h} y_{h_i} = \frac{Y}{N} = \bar{Y},$$

the population mean of all the N values of y in the population. Thus, $\bar{y}_{st}$ is an unbiased estimator of the population mean.

Moreover,

$$V(\bar{y}_{st}) = \frac{1}{N^2} \sum_{h=1}^{H} N_h^2 V(\bar{y}_h)$$

$$= \sum_{h=1}^{H} W_h^2 \left(\frac{1}{n_h} - \frac{1}{N_h}\right) S_h^2$$

writing $W_h = \frac{N_h}{N}$.

For an SRSWOR of size $n = \sum_{1}^{H} n_h$ taken from the entire population of $N =$

$\sum_{h=1}^{H} N_h$ units the sample mean $\bar{y} = \frac{1}{n} \sum_{j \in s} y_j$ writing s for the SRSWOR of size n from

$U = (1, \ldots, j, \ldots, N)$, we know we have

$$V(\bar{y}) = \frac{N-n}{Nn} \frac{1}{N-1} \sum_{j=1}^{N} (y_j - \bar{Y})^2$$

$$= \left(\frac{1}{n} - \frac{1}{N}\right) S^2 \text{ writing}$$

$$S^2 = \frac{1}{N-1} \sum_{j=1}^{N} (y_j - \bar{Y})^2 .$$

However, we may write

$$\sum_{j=1}^{N} (y_j - \bar{Y})^2 = \sum_{h=1}^{H} \sum_{i=1}^{N_h} (y_{h_i} - \bar{Y})^2$$

$$= \sum_{h=1}^{H} \sum_{i=1}^{N_h} \left\{(y_{h_i} - \bar{Y}_h) + (\bar{Y}_h - \bar{Y})\right\}^2$$

$$= \sum_{h=1}^{H} \sum_{i=1}^{N_h} (y_{h_i} - \bar{Y}_h)^2 + \sum_{h=1}^{H} N_h (\bar{Y}_h - \bar{Y})^2$$

$$+ 2 \sum_{h=1}^{H} (\bar{Y}_h - \bar{Y}) \sum_{i=1}^{N_h} (y_h - \bar{Y}_h)$$

$$= \sum_{h=1}^{H} \sum_{i=1}^{N_h} (y_{h_i} - \bar{Y}_h)^2 + \sum_{h=1}^{H} N_h (\bar{Y}_h - Y)^2 .$$

It is customary to regard $\sum_{j=1}^{N} (y_j - \bar{Y})^2$ as the Total Sum of Square (TSS) in the

populations, $\sum_{h=1}^{H} N_h (\bar{Y}_h - \bar{Y})^2$ as the Between Sum of Squares (BSS) among the

strata means and

$$\sum_{h=1}^{H} \sum_{i=1}^{N_h} (y_{h_i} - \bar{Y}_h)^2 = \sum_{h=1}^{H} (N_h - 1) S_h^2$$

as the Within Sum of Squares (WSS) among the strata values. Thus,

$$TSS = BSS + WSS.$$

Now, $$V(\bar{y}) = \left(\frac{1}{n} - \frac{1}{N}\right) S^2 = \left(\frac{1}{n} - \frac{1}{N}\right)(N-1)\, TSS$$

is determined by the TSS but $V(\bar{y}_{st}) = \sum_{h=1}^{H} W_h^2 \left(\frac{1}{n_h} - \frac{1}{N_h}\right) S_h^2$ is a function of the within sum of squares WSS. In $V(\bar{y})$, both BSS and WSS are involved but $V(\bar{y}_{st})$ is bereft of the BSS component altogether in TSS. So, $V(\bar{y}_{st})$ is expected to be less than $V(\bar{y})$ if BSS is a sizeable component in TSS. Thus, follows the Principle: Stratification is effective if the strata are so constructed that the strata are internally as homogeneous as possible and the strata themselves in terms of the strata means as much divergent as possible.

Since from every stratum an SRSWOR is drawn, an unbiased estimator of S_h^2 is

$$s_h^2 = \frac{1}{(n_h - 1)} \sum_{i \in s_h} \left(y_{h_i} - \bar{Y}_h\right)^2.$$

So, an unbiased estimator of $V(\bar{y}_{st})$ is

$$\nu(\bar{y}_{st}) = \sum_{h=1}^{H} W_h^2 \left(\frac{1}{n_h} - \frac{1}{N_h}\right) s_h^2.$$

If an SRSWOR of size $n = \sum_{h=1}^{H} n_h$ was drawn from the population and the sample mean $\bar{y}$ was used to unbiasedly estimate the population mean $\bar{Y}$, then we would get

$$V(\bar{y}) = \frac{N-n}{Nn} S^2 = \frac{N-n}{Nn} \frac{1}{N-1} \sum_{1}^{N} \left(y_i - \bar{Y}\right)^2$$

$$= \left(\frac{1}{n} - \frac{1}{N}\right) \frac{1}{N-1} \left[\sum_{h=1}^{H} \sum_{i=1}^{N_h} y_{h_i}^2 - N(\bar{Y})^2\right].$$

Now, $$V(\bar{y}_{st}) = E(\bar{y}_{st})^2 - (\bar{Y})^2.$$

So, from a stratified SRSWOR, we could obtain an unbiased estimator for $(\bar{Y})^2$ as

$$\nu_{st}\left[(\bar{Y})^2\right] = (\bar{y}_{st})^2 - \nu(\bar{y}_{st}).$$

So, from the same stratified SRSWOR, we could unbiasedly estimate $V(\bar{y})$ by

$$\nu_{st}(\bar{y}) = \frac{\left(\frac{1}{n} - \frac{1}{N}\right)}{N-1} \left[\sum_{h=1}^{H} \frac{N_h}{n_h} \sum_{i \in s_h} y_{h_i}^2 - N \left\{ (\bar{y}_{st})^2 - \nu(\bar{y}_{st}) \right\} \right]$$

Then, $\delta = [\nu_{st}(\bar{y}) - \nu(\bar{y}_{st})]$ may be taken as an unbiased measure of the gain in efficiency in estimating $\bar{Y}$ by stratified SRSWOR rather than by an SRSWOR with a common sample-size $n = \sum_{h=1}^{H} n_h$; the higher the positive magnitude of δ, the higher the gain. Of course, a negative value of δ will indicate a 'loss' rather than a 'gain' in efficiency by dint of stratification. A 'loss' is rather unlikely to encounter in practice.

3.2 Allocation of Sample-Size

Since,
$$V(\bar{y}_{st}) = \sum_{h=1}^{H} W_h^2 \left(\frac{1}{n_h} - \frac{1}{N_h} \right) S_h^2$$

$$= \frac{1}{N^2} \left[\sum_{h=1}^{H} \frac{N_h^2 S_h^2}{n_h} \right] - \frac{1}{N^2} \sum N_h S_h^2$$

we may examine how a rationale may emerge about how to settle a question about assigning the sample-sizes $n_h (h = 1, \cdots, H)$ stratum-wise, keeping in mind the magnitudes of $n = \sum_{h=1}^{H} n_h$, N_h and S_h for $h = 1, \ldots, H$.

Cauchy inequality tells us

$$\left(\sum_{h=1}^{H} n_h \right) \left(\sum_{h=1}^{H} \frac{N_h^2 S_h^2}{n_h} \right) \geq \left(\sum_{h=1}^{H} N_h S_h \right)^2 \qquad (3.2.1)$$

implying the choice
$$n_h \propto N_h S_h$$

giving the minimal value of the quantity on the left of (3.2.1). Hence, the allocation formula
$$n_h = \frac{n N_h S_h}{\sum_{h=1}^{H} N_h S_h}$$

originally given by Neyman (1934) is an appropriate sample allocation formula in stratified SRSWOR.

Since S_h-values in practice cannot be known, the formula

$$n_h = n\frac{N_h}{N}$$

ignoring S_hs in Neyman's allocation formula is often employed in practice and is called the 'Proportional sample allocation' rule as a by-product from the formula.

A more complicated 'allocation formula' is also available as follows in the literature.

Let $C_h, h = 1, \ldots, H$ denote the stratum-specific cost of sample selection and sample survey works per unit of sampling and C_0 denote an over-head cost of survey. Then,

$$C = C_0 + \sum_{h=1}^{H} C_h n_h$$

may be taken as the over-all survey cost for $n = \sum_{h=1}^{H} n_h$ sampled units. On the other

hand, $V(\bar{y}_{st}) = \frac{1}{N^2}\sum_{h=1}^{H}\frac{N_h^2 S_h^2}{n_h} - \frac{1}{N^2}\sum_{h=1}^{H} N_h S_h^2 = V$, say, is the variance of $\bar{y}_{st}$. A

possible 'allocation' approach may be to consider the product

$$(C - C_0)\left(V + \frac{1}{N^2}\sum_{h=1}^{H} N_h S_h^2\right),$$

minimize it with respect to n_h and settle for its choice with either fixing V or C supposing the other has been minimized.

The Cauchy Inequality is then to be considered as

$$\left(\sum_{h=1}^{H} C_h n_h\right)\left(\sum_{h=1}^{H} W_h^2\frac{S_h^2}{n_h}\right) \geq \left(\sum W_h S_h \sqrt{C_h}\right)^2. \tag{3.2.2}$$

The quantity on the left of (3.2.2) is minimized for the choice

$$n_h \propto \frac{W_h S_h}{\sqrt{C_h}}$$

leading to the rational allocation rule

$$n_h = \frac{n\frac{W_h S_h}{\sqrt{C_h}}}{\sum\limits_{h=1}^{H} \frac{W_h S_h}{\sqrt{C_h}}}, h = 1, \ldots, H.$$

This is more general than Neyman's allocation rule and is known as the 'optimal sample-size allocation' formula in stratified SRSWOR.

It is apparently logical because the higher the values of N_h and S_h and the smaller the stratum-wise operational cost, the higher the stratum-specific sample-size one should be prepared to afford for a good survey result.

3.3 Construction of Strata

Dalenius (1950) gave us a theoretical solution to the problem of constructing appropriately the strata in a survey sampling problem. Unlike the way we have treated the survey sampling approach thus far in this book, his discussion is too theoretically inclined.

Let the population values of y may be such that the units $i(= 1, \ldots, N)$ of the population $U = (1, \ldots, i, \ldots, N)$ with values y_i constitute the hth stratum such that

$$a_{h-1} < y_i \leq a_h,$$

when
$$-\infty < a_0 < a_1 < \cdots, < a_{h-1} < a_h$$
$$< a_{h+1} < \cdots, < a_{H-1} < a_H < +\infty$$

and let y have the probability density function $f(.)$ and the cumulative probability distribution function $F(.)$ such that

$$F(x) = \int_{-\infty}^{x} f(t)dt; \quad \frac{dF(x)}{dx} = f(x)$$

Let
$$W_h = \int_{a_{h-1}}^{a_h} f(t)dt$$
$$= F(a_h) - F(a_{h-1}) = \text{Prob}(a_{h-1} < y \leq a_h)$$

Then
$$\frac{\partial W_h}{\partial a_h} = f(a_h);$$

$$\mu_h = \int_{a_{h-1}}^{a_h} tf(t)dt \Big/ \int_{a_{h-1}}^{a_h} f(t)dt$$

$$\equiv \text{ the } h^{th} \text{ stratum mean of } y.$$

$$S_h^2 = \frac{\int_{a_{h-1}}^{a_h} t^2 f(t)\,dt}{\int_{a_{h-1}}^{a_h} f(t)\,dt} - \left[\frac{\int_{a_{h-1}}^{a_h} t F(t)\,dt}{\int_{a_{h-1}}^{a_h} f(t)\,dt}\right]^2$$

$$W_h S_h^2 = \int_{a_{h-1}}^{a_h} t^2 f(t)\,dt - \frac{\left[\int_{a_{h-1}}^{a_h} t f(t)\,dt\right]^2}{\int_{a_{h-1}}^{a_h} f(t)\,dt}$$

So,

$$\frac{\partial W_h}{\partial a_h} S_h^2 = S_h^2 \frac{\partial W_h}{\partial a_h} + 2 W_h S_h \frac{\partial S_h}{\partial a_h}$$

$$= a_h^2 f(a_h) - 2 a_h f(a_h)\mu_h + \mu_h^2 f(a_h)$$

Now,

$$S_h^2 \frac{\partial W_h}{\partial a_h} = S_h^2 f(a_h).$$

So,

$$2 S_h^2 \frac{\partial W_h}{\partial a_h} + 2 S_h W_h \frac{\partial S_h}{\partial a_h}$$

$$= a_h^2 f(a_h) + S_h^2 f(a_h) - 2 a_h f(a_h)\mu_h + \mu_h^2 f(a_h)$$

or,

$$S_h \frac{\partial W_h}{\partial a_h} + W_h \frac{\partial S_h}{\partial a_h}$$

$$= \frac{a_h^2 f(a_h)}{2 S_h} + \frac{S_h^2 f(a_h)}{2 S_h} - \frac{a_h f(a_h)\mu_h}{S_h} + \frac{\mu_h^2 f(a_h)}{2 S_h}$$

$$= \frac{1}{2}\frac{f(a_h)}{S_h}\left[a_h^2 - 2 a_h \mu_h + \mu_h^2 + S_h^2\right]$$

$$= \frac{1}{2}\frac{f(a_h)}{S_h}\left[(a_h - \mu_h)^2 + S_h^2\right]$$

or,

$$\frac{\partial(W_h S_h)}{\partial a_h} = \frac{1}{2}\frac{f(a_h)}{S_h}\left[(a_h - \mu_h)^2 + S_h^2\right]$$

Similarly, one may derive

$$\frac{\partial}{\partial a_h}(W_{h+1} S_{h+1}) = \frac{1}{2}\frac{f(a_h)}{S_{h+1}}\left[(a_h - \mu_{h+1})^2 + S_{h+1}^2\right].$$

Let us now recall that

$$V(\bar{y}_{st}) = \frac{1}{N^2}\sum N_h^2\left(\frac{1}{n_h} - \frac{1}{N_h}\right)S_h^2$$

$$= \sum_h W_h^2 \frac{S_h^2}{n_h}, \text{ neglecting } \frac{n_h}{N_h}.$$

Also, Neyman's formula gives us the

$$V(\bar{y}_{st})\text{Neyman} = \frac{1}{n}\left(\sum_h W_h S_h\right)^2.$$

So,
$$\frac{\partial}{\partial a_h}\left[V(\bar{y}_{st})\text{Neyman}\right] = 0$$

is required to be solved to minimize $V(\bar{y}_{st})\text{Neyman}$ to optimally choose a_h (for $h = 1, \ldots, N$).

So, we need to solve $\frac{\partial}{\partial a_h}\sum_{h=1}^{H} W_h S_h = 0$, i.e., we need to solve

$$\frac{\partial}{\partial a_h}(W_h S_h + W_{h+1}S_{h+1}) = 0$$

because other terms in $\frac{\partial}{\partial a_h}\left(\sum_{h=1}^{H} W_h S_h\right)$ are free of a_h.

So, an optimal a_h is to be found from

$$\frac{(a_h - \mu_h)^2 + S_h^2}{S_h} = -\frac{(a_h - \mu_{h+1})^2 + S_{h+1}^2}{S_{h+1}}.$$

That is hard to solve because $\mu_h, S_h, \mu_{h+1}, S_{h+1}$ involve a_h each.

So, Dalenius (1950) could not solve the problem to choose a_h optimally. But Dalenius himself and many of his followers like Hodges, Gurney and many others pursued with this problem producing many more theoretical results, too tough to be presented here any further.

3.4 Post-Stratification

A situation may be visualized when the strata may be defined such that the interval

$$(a_{h-1}, a_h], h = 1, \ldots, H$$

may be defined with $-\infty < a_0 < a_1 < \cdots < a_{h-1} < a_h < \cdots < a_{H-1} < a_H < +\infty$ so that the y-values with $a_{h-1} < y \leq a_h$ denoting the hth stratum, $h = 1, \cdots, H$.

But the units belonging to the respective H strata may not be identifiable even though the strata-sizes $N_h(N_1, \ldots, N_h, \ldots, N_H)$ and strata proportions may be known prior to an actual survey. Moreover, while implementing the survey, it may still be possible to assign the individuals with respective y-values measured, to the pre-defined strata

$$(a_{h-1}, a_h \,], h = 1, \ldots, H.$$

In such a pre-planned situation, in order to effectively estimate the population mean

$$\bar{Y} = \frac{1}{N} \sum_{i=1}^{N} y_i,$$

one may take an SRSWOR s of a pre-assigned size n but after the survey may use instead of the sample mean $\bar{y} = \frac{1}{n} \sum_{i \in s} y_i$ a possibly better estimate described below.

Let $n_h(h = 1, \ldots, H)$ be the observed numbers of units in s found to have y-values in the respective intervals

$$J_h = (a_{h-1}, a_h \,] \text{ for } h = 1, \ldots, H.$$

Let

$$I_h = 1, \text{ if } n_h > 0$$
$$= 0, \text{ if } n_h = 0$$

Thus,

$$E(I_h) = \text{Prob}(n_h > 0)$$
$$= 1 - \binom{N - N_h}{n} \Big/ \binom{N}{n}$$

implying the probability that all the observed n sampled units have not come from outside the interval J_h, i.e., some units at least come with y-values in J_h.

Let

$$\bar{y}'_h = \frac{1}{n_h} \sum_{i \in J_h} y_i \text{ if } n_h > 0$$
$$= 0 \text{ if } n_h = 0, n = 1, \cdots, N$$

Then

$$\bar{y}'\text{pst} = \sum_{h=1}^{H} W_h \bar{y}_{h'} \frac{I_h}{E(I_h)}$$

may be taken as an unbiased estimator for $\sum_{h=1}^{H} W_h \bar{Y}_h = \bar{Y}$. This $\bar{y}'_{pst}$ is called an unbiased post-stratified estimator for $\bar{Y}$ and the intervals $J_h = (a_{h-1}, a_h]$ are called the post-strata because only posterior to the completion of an implemented survey of an SRSWOR of size n one ascertains the n_h units falling in J_h for $h = 1, \ldots, H$.

Doss et al. (1979) recovered the following ratio-like biased post-stratified estimator for $\bar{Y}$, namely

$$\bar{y}''_{pst} = \frac{\sum_{h=1}^{H} W_h \bar{y}_{h'} \frac{I_h}{E(I_h)}}{\sum_{h=1}^{H} W_h \frac{I_h}{E(I_h)}}.$$

This, though not unbiased for $\bar{Y}$ because it satisfies the 'linear invariance' property in the sense that if $y_i = a + bx_i$ for certain constants a, b and given values of x_is one may check that

$$\bar{y}''_{pst}(y) = a + b\bar{x}''_{pst}$$

using (y), (x) to emphasize that the estimators, respectively, are $\bar{y}''_{pst}(.)$ with y and x in $(.)$. This property is not shared by $\bar{y}'_{pst}$ as is quite easy to check.

3.5 Sample Selection with Probability Proportional to Size with Replacement and Its Various Amendments

In SRSWR and SRSWOR, there is no conscious effort to enhance the efficiency in estimation except through increasing the number of draws and the sample-size, respectively.

In stratified sampling, a conscious effort is to achieve an improved efficiency by dint of construction of strata achieving substantial within strata control in variability. Another point to note is that in stratified SRSWOR inclusion probability of a unit falling in the hth stratum is

$$\frac{n_h}{N_h} \text{ for } h = 1, \ldots, H.$$

So, it is 'unequal probability sampling' unless

$$\frac{n_h}{N_h} = \frac{n}{N} \text{ for every } h = 1, \ldots, H.$$

So, 'varying probability sampling' or 'unequal probability sampling' rather than 'equal probability' opens out a possibility of endeavouring to achieve higher efficiency in estimation in appropriate ways, without adjusting the sample-size alone.

3.5.1 'Probability proportional to size' or PPS Method of Sample Selection is an Attempt Towards Enhancing Efficiency in Estimation

Suppose we are interested in estimating the population total

$$Y = \sum_1^N y_i$$

of a variable y valued y_i for the ith unit of the population $U = (1, \ldots, i, \ldots, N)$. Let, however, the values x_i of an auxiliary variable x be known for every unit i of U having a total $X = \sum_1^N x_i$ and a mean $\bar{X} = \frac{X}{N}$, and moreover, every x_i be positive. If x and y may be supposed to be positively correlated, then one may use the x_i-values in the sample selection in the following way and use the chosen sample in producing a new unbiased estimator for Y with a clearly specified variance formula for an unbiased estimator of Y admitting an unbiased variance estimator as well with a simple form.

PPS Selection with Replacement (PPSWR)

Let us tabulate the x_i-values.

Unit (i)	x_i	Cumulative x_i s C_i
1	x_1	$C_1 = x_1$
2	x_2	$C_2 = x_1 + x_2$
3	x_3	$C_3 = x_1 + x_2 + x_3$
$\vdots$	$\vdots$	$\vdots$
i	x_i	$C_i = x_1 + x_2 + \cdots + x_i$
$\vdots$	$\vdots$	$\vdots$
N	x_N	$C_N = x_1 + x_2 + \cdots + \cdots + x_N = X$

For simplicity, let every x_i be a positive integer; if originally they were not so, each can be made so multiplying each by a number 10^r, with r taken as a suitable positive integer, say, 2, 3, 4, etc. This adjustment will only help selection without adversely affecting unbiasedness and efficiency in estimation.

Let a random number R be chosen between 1 and X. If it is found that

$$C_{i-1} < R \leq C_i, i = 1, \ldots, N$$

with $C_0 = 0$, then the unit i is to be sampled. Then the selection probability of i is

$$p_i = \frac{C_i - C_{i-1}}{X} = \frac{x_i}{X}, i = 1, \ldots, N.$$

Thus, by this method of selecting the unit i is selected with a probability proportional to its size

$$p_i = \frac{x_i}{X}, i \in U$$

regarding x_i as the size-measure of the unit i and calling x as a size-measure variable associated with the units of U, called the sampling units, that is, the set of units of U out of which a selection is to be made. This is called the Cumulative PPS method. Let us turn to an alternative procedure to select a unit of U with a probability proportional to size or PPS, given to us by Lahiri (1951). Let as before every x_i be made a positive integer if not already so.

Given the vector $\underline{X} = (x_1, \ldots, x_i, \ldots, x_N)$ with each $x_i > 0$ for i in U, let M be a positive integer such that $M \geq x_i$ for every i in U.

Lahiri's (1951) PPS method demands choosing a unit i, say, of U with a probability $\frac{1}{N}$, i.e., at random and choosing another random number R between 1 and M.

Then, if $R \leq x_i$, then take it in the sample and else drop the selection (i, R) and repeat this procedure again and again till a unit happens to get selected by the above rule. Lahiri (1951) proves as follows that after a finite number of such repetition a unit happens to get selected and the probability that i is selected turns to be

$$p_i = \frac{x_i}{X}, i \in U.$$

Let us verify this below.

Following through the manner in which a unit, say, i is to be selected, let us easily check the probability that a unit i of U may be selected turns out as

$$P(i) = \text{Prob}(i \text{ is selected in a sample})$$
$$= \frac{1}{N}\frac{x_i}{M} + Q\left(\frac{1}{N}\frac{x_i}{M}\right) + Q^2\left(\frac{1}{N}\frac{x_i}{M}\right)$$
$$+ \cdots + Q^r\left(\frac{1}{N}\frac{x_i}{M}\right) + \cdots \tag{3.5.1}$$

with r as a positive integer and

$$Q = 1 - \frac{1}{N}\sum_{i=1}^{N}\frac{x_i}{M} = 1 - \frac{\bar{X}}{M} \tag{3.5.2}$$

with $0 < Q < 1$ by definition of $\bar{X}$ and M.

Because of (3.5.2), it follows that (3.5.1) gives us the result

$$P(i) = \frac{1}{N} \frac{x_i}{M} \left(1 + Q + Q^2 + \cdots Q^r + \cdots + \infty\right)$$

$$= \frac{1}{N} \frac{x_i}{M} (1 - Q)^{-1} = \frac{1}{N} \frac{x_i}{M} \frac{M}{\bar{X}}$$

$$= \frac{x_i}{X} = p_i.$$

Since the series $\left(1 + Q + Q^2 + \cdots\right)$ is covergent, clearly the step in Lahiri's (1951) method continues at most a finite number of times and the selection probability of i in a sample is

$$p_i = \frac{x_i}{X}.$$

Thus, by this method, as in the cumulative PPS method, also a PPS sampled unit emerges of course.

The sampling scheme that repeats independently a number of times n, say, either the 'cumulative total' or 'Lahiri's (1951) PPS sampling', is called PPSWR, i.e., 'Probability Proportional to size with Replacement' scheme of sample selection.

Hansen and Hurwitz (1943) have given us the following method of unbiasedly estimating the 'Finite Population Total' $Y = \sum_{i=1}^{N} y_i$, namely the Hansen and Hurwitz estimator

$$t_{HH} = \frac{1}{n} \sum_{r=1}^{n} \frac{y_r}{p_r},$$

writing y_r, p_r as the values of y_i, p_i for the unit i chosen on the rth $(r = 1, \ldots, n)$ draw by the PPS method of selection.

For every draw $r(= 1, \ldots, n)$, we have

$$E\left(\frac{y_r}{p_r}\right) = \sum_{i=1}^{N} \frac{y_i}{p_i} \mathrm{Prob}\left(\frac{y_r}{p_r} = \frac{y_i}{p_i}\right)$$

$$= \sum_{i=1}^{N} \frac{y_i}{p_i} p_i = Y \text{ for every } r(= 1, \ldots, n).$$

So,
$$E(t_{HH}) = \frac{1}{n} \sum_{r=1}^{n} E\left(\frac{y_r}{p_r}\right) = Y$$

$$V(t_{HH}) = \frac{1}{n^2} \sum_{r=1}^{n} V\left(\frac{y_r}{p_r}\right)$$

$$V\left(\frac{y_r}{p_r}\right) = E\left(\frac{y_r}{p_r} - Y\right)^2$$

$$= \sum_{i=1}^{N} \left(\frac{y_i}{p_i} - Y \right)^2 \text{Prob} \left(\frac{y_r}{p_r} = \frac{y_i}{p_i} \right)$$

$$= \sum_{i=1}^{N} p_i \left(\frac{y_i}{p_i} - Y \right)^2 \tag{i}$$

$$= \sum_{i=1}^{N} \frac{y_i^2}{p_i} - Y^2 \tag{ii}$$

$$= \sum_{i<j}^{N} \sum^{N} p_i p_j \left(\frac{y_i}{p_i} - \frac{y_j}{p_j} \right)^2 \tag{iii}$$

$$= V, \text{ say, for every } r = 1, \dots, n$$

So, $$V(t_{HH}) = \frac{V}{n} = \frac{1}{n} \sum_{i<j} \sum p_i p_j \left(\frac{y_i}{p_i} - \frac{y_j}{p_j} \right)^2$$

Also, $$E \left(\frac{y_r}{p_r} - \frac{y_r'}{p_r'} \right)^2 \text{ for every } r \neq r' (= 1, \dots, n)$$

equals $$E \left[\left(\frac{y_r}{p_r} - Y \right) - \left(\frac{y_r'}{p_r'} - Y \right) \right]^2$$

$$= E \left(\frac{y_r}{p_r} - Y \right)^2 + E \left(\frac{y_r'}{p_r'} - Y \right)^2 = 2V$$

So, $$\nu = \frac{1}{2n^2(n-1)} \sum_{r \neq r'}^{n} \sum^{n} \left(\frac{y_r}{p_r} - \frac{y_r'}{p_r'} \right)^2$$

has $$E(\nu) = \frac{1}{2n^2(n-1)} (2V) n \times (n-1) = \frac{V}{n} = V(t_{HH}).$$

Hence, ν is an unbiased estimator for $V(t_{HH})$.

If $x_i = 1$ for every $i = 1, \cdots, N$, then $p_i = \frac{1}{N}$ for every $i = 1, \dots, N$, and then (A) PPSWR sampling reduces to SRSWR, (B) t_{HH} reduces to $N\bar{y}$ and (C) $V(t_{HH})$ reduces to

$$\frac{N}{n} \sum_{1}^{N} (y_i - \bar{Y})^2 \text{ (vide (i) above)}$$

$$= \frac{N^2}{n} \sigma^2, \text{ since } \sigma^2 = \frac{1}{N} \sum_{1}^{N} (y_i - \bar{Y})^2.$$

With x_is at hand, we have a flexibility to resort to PPSWR sampling rather than to SRSWR enabling us to employ t_{HH} rather than $N\bar{y}$ achieving a variance $V(t_{HH}) = \frac{V}{n}$ rather than $V(N\bar{y}) = N^2 \frac{\sigma^2}{n}$. However, there is no guarantee that

$$\frac{V}{n} < \frac{N^2\sigma^2}{n},$$

$$\text{i.e.,} \sum_1^N p_i \left(\frac{y_i}{p_i} - Y\right)^2 < N \sum_1^N (y_i - \bar{Y})^2. \tag{iv}$$

But if y is well and positively correlated with x, we may hopefully expect that (iv) may hold true. To examine its plausibility, let us postulate a helpful model so as to write

$$y_i = \beta x_i + \epsilon_i, i \in U$$

with β as an unknown constant, and ϵ_is are random variables that are independent with means $E_m(\epsilon_i) = 0$ and variances $V_m(\epsilon_i) = \tau^2$ for every i in U.

The relation (iv) may also be written as

$$V = \sum\sum_{i<j} p_i p_j \left(\frac{y_i}{p_i} - \frac{y_j}{p_j}\right)^2 < \sum\sum_{i<j}(y_i - y_j)^2. \tag{v}$$

But we may easily check that

$$E_m(V) = \tau^2 \left(X \sum_1^N \frac{1}{x_i} - N\right) \tag{vi}$$

and also, clearly

$$E_m \sum\sum_{i<j}(y_i - y_j)^2 = N\beta^2 \sum_1^N (x_i - \bar{X})^2 + \tau^2 N(N - 1). \tag{vii}$$

Further, since for positive numbers their Arithmetic mean A cannot be smaller than their Harmonic mean H, it follows that $E_m(V) > \tau^2 N(N - 1)$.

So, we cannot say that generally (vii) may exceed (vi). But if every x_i is the same, (vi) just equals (vii). So, whether PPSWR may yield a better unbiased estimator than SRSWR remains an open question. But since the x_is are all at hand, it us easy to check the magnitudes of

$$X \sum_1^N \frac{1}{x_i} - N \text{ and of } \sum_1^N (x_i - \bar{X})^2$$

and thus tabulate the values of $E_m(V)$ and $E_m N \sum_1^N (y_i - \bar{Y})$ for various choices of (β, τ) in order to recommend PPSWR as a Competitor versus SRSWR.

For PPSWR in n draws, the inclusion probability of i is

$$\pi_i = 1 - (1 - p_i)^n, i \in U \text{ and}$$

that of i and $j (i \neq j)$ is

$$\pi_{ij} = 1 - (1 - p_i)^n - (1 - p_j)^n + (1 - p_i - p_j)^n.$$

Suppose a PPSWR sample taken in n draws has been surveyed and t_{HH} and $v =$ an unbiased estimator of $V(t_{HH})$ have been calculated. Then one may as follows unbiasedly estimate how much additional accuracy has been realized vis-a-vis what might have been achieved if instead an SRSWR in n draws had been taken and $N\bar{y}$ was used to unbiasedly estimate Y.

Let us note

$$t_{HH} = \frac{1}{n} \sum_{r=1}^{n} \frac{y_r}{p_r}$$

$$v = \frac{1}{2n^2(n-1)} \sum_{r \neq r'} \sum \left(\frac{y_r}{p_r} - \frac{y'_r}{p'_r} \right)^2 = v(t_{HH}), \text{ say}$$

$$V(t_{HH}) = E(t_{HH}^2) - Y^2$$

$$\bar{y} = \frac{1}{n} \sum_{r=1}^{n} y_r$$

$$\begin{matrix} V(\bar{y}) \\ \text{SRSWR}(n) \end{matrix} = \frac{1}{Nn} \sum_{r=1}^{n} (y_r - \bar{Y})^2 = E(\bar{y}^2) - (\bar{Y})^2$$

$$\begin{matrix} V(N\bar{y}) \\ \text{SRSWOR}(n) \end{matrix} = E(N\bar{y})^2 - Y^2.$$

$$\delta = \frac{V(N\bar{y})}{\text{SRSWR}(n)} - V(t_{HH}) \text{ is to be unbiasedly estimated from a PPSWR}(n) \text{ sample}$$
by

$$\delta = \frac{N^2}{n^2} \left[\sum_{i \in s} \frac{y_i^2}{\pi_i} + \sum_{i \neq j \in s} \sum \frac{y_i y_j}{\pi_{ij}} \right] - v(t_{HH})$$

writing s as the sample of distinct units observed in the PPSWR in n draws and

$$\pi_i = 1 - (1 - p_i)^n \text{ and } \pi_{ij} = 1 - (1 - p_i)^n - (1 - p_j)^n + (1 - p_i - p_j)^n.$$

A positive and high-valued δ will indicate superior accuracy of t_{HH} based on PPSWR over $N\bar{y}$ from SRSWR.

3.5.2 *Probability Proportional to Size without Replacement Sampling (PPSWOR) and Estimation*

Des Raj (1956) gave the following method of sample selection with probability proportional to size (PPS) without replacement (PPSWOR) and the related estimation procedures.

Let as in PPSWR sampling positive integers x_i be available as numbers associated with every unit i of $U = (1, \ldots, i, \ldots, N)$ with $X = \sum_1^N x_i$ and $p_i = \frac{x_i}{X}$ be the normed size-measures. Either by the cumulative total method or by Lahiri's (1951) method let one unit, say, i be selected by the PPS method on the first draw. After setting aside the selected unit from the population, let from the remainder of the population $U - \{i\}$, one unit, say, $j(\neq i)$ be similarly chosen by the PPS method with a probability proportional to

$$Q_j = \frac{p_j}{1 - p_i}, \, j \neq i, (1, \ldots, \neq i, \ldots, N)$$

by the cumulative total or Lahiri's (1951) using these adjusted normed size-measures Q_j as above as are now relevant, then $\sum_{j \neq i} Q_j = 1$.

Let this procedure be repeated till $n(> 2)$ distinct units be chosen from U.

Des Raj (1956) observes that the following n unbiased estimators for $Y = \sum_1^N y_i$ are now available on recording that $i_1, i_2, \ldots, i_j, \ldots, i_n$ are the n distinct units thus chosen from U with respective y-values as $y_{i_1}, y_{i_2}, \ldots, y_{i_j}, \ldots, y_{i_n}$. Then (Des Raj, 1956) notes that

$$t_1 = \frac{y_{i_1}}{p_{i_1}}, t_2 = y_{i_1} + \frac{y_{i_2}}{p_{i_2}}(1 - p_{i_1}), \ldots$$

$$t_j = y_{i_1} + \cdots + y_{i_{j-1}} + \frac{y_{i_j}}{p_{i_j}}(1 - p_{i_1}, \ldots, p_{i_{j-1}})$$

$$\vdots$$

$$t_n = y_{i_1} + \cdots, + y_{i_{n-1}} + \frac{y_{i_n}}{p_{i_n}}(1 - p_{i_1}, \cdots p_{i_{n-1}})$$

are the n estimators for Y each unbiased.

Let us easily check the following:

$$E(t_1) = \sum_1^N \frac{y_i}{p_i} = Y$$

$$E(t_2) = E\left[E(t_2)|(i_1, y_{i_1})\right]$$

$$= E\left[y_{i_1} + \sum_{\substack{j=1 \\ \neq i_1}}^N \frac{y_j}{p_j}\frac{p_j}{(1 - p_{i_1})}\right]$$

$$= E\left[y_{i_1} + (Y - y_{i_1})\right] = E(Y) = Y,$$

$$E(t_j) = EE\left[t_j | (i_1, y_{i_1}), \dots, (i_{j-1}, y_{i_{j-1}})\right]$$

$$= E\left[(y_{i_1} + \dots + y_{i_{j-1}}) + \sum_{\substack{K=1 \\ (\neq i_1, \dots, i_{j-1})}}^{N} \frac{y_K}{p_K} \frac{p_K(1 - p_{i_1} - \dots - p_{i_{j-1}})}{(1 - p_{i_1} - \dots - p_{i_{j-1}})}\right]$$

$$= E\left[(y_{i_1} + \dots + y_{i_{j-1}}) + \sum_{\substack{K=1 \\ (\neq i_1, \dots, i_{j-1})}}^{N} y_K\right]$$

$$= E(Y) = Y.$$

Similarly, $E(t_j) = Y$ too.

So, Des Raj (1956) recommends for Y the unbiased estimator

$$t_D = \bar{t} = \frac{1}{n} \sum_{j=1}^{n} t_j \text{ since } E(t_D) = E(\bar{t}) = Y.$$

Now, for any $j' > j, j = 1, \dots, n - 1,$

$$E(t_j t_{j'}) = E(t_j) E(t_{j'})(i_1 \cdots i_j, y_{i_1} \cdots y_{i_j})]$$
$$= E(t_j) Y = Y^2.$$

So, $\text{Cov}(t_j, t_{j'}] = 0$ for every $j' > j$ and $j = 1, \dots, n - 1$.

So, $V(t_D) = V\left(\frac{1}{n} \sum_{1}^{n} t_j\right) = \frac{1}{n^2} \sum_{j=1}^{n} V(t_j).$

As a result, an unbiased estimator for $V(t_D)$ is

$$v(t_D) = \frac{1}{2n^2(n-1)} \sum_{\substack{j \neq j'}}^{n} \sum^{n} (t_j - t_{j'})^2$$

because,

$$E(t_j - t_{j'})^2 = E\left[(t_j - Y) - (t_{j'} - Y)\right]^2$$
$$= V(t_j) + V(t_{j'})$$

and

$$Ev(t_D) = \frac{1}{2n^2(n-1)} \left[2(n-1) \sum_{1}^{n} V(t_j)\right]$$

$$\frac{1}{n^2} \sum_{j=1}^{n} V(t_j) = V(\bar{t}) = V(t_D)$$

Let us further note the following:

$$V_1 = V(t_1) = E\left(\frac{y_{i_1}}{p_{i_1}} - Y\right)^2 = \sum_{i=1}^{N} p_i \left(\frac{y_i}{p_i} - Y\right)^2$$

$$= V = V(t_{HH}) \text{ when } n = 1,$$

$$V_2 = V(t_2) = V\left[y_{i_1} + \frac{y_{i_2}}{p_{i_2}}(1 - p_{i_1})\right]^2$$

$$= E\left[V(t_2 | p_{i_1}, y_{i_1})\right]$$

$$= E\sum_{\substack{j<K \\ \neq i}}^{N}\sum^{N} \left(\frac{p_j}{1 - p_i}\middle|\frac{p_K}{1 - p_i}\right)\left\{\frac{y_j}{p_j}(1 - p_i) - \frac{y_K}{p_K}(1 - p_i)\right\}^2$$

$$= E\sum_{\substack{j<K \\ (\neq i)}}\sum p_j p_K \left(\frac{y_j}{p_j} - \frac{y_K}{p_K}\right)^2$$

$$= \sum_{j<K}\sum (1 - p_j - p_K) p_j p_K \left(\frac{y_j}{p_j} - \frac{y_K}{p_K}\right)^2$$

$$< V = V_1 \text{ because } 0 < (1 - p_j - p_K) < 1.$$

For simplicity, let

$$Q_j = \frac{p_j}{1 - p_i}, \, j = 1, \ldots, N(\neq i),$$

$$P_K = \frac{Q_K}{1 - Q_i}, \, K = 1, \ldots, N(\neq i, j).$$

Similarly,

$$V_3 = V(t_3) = \sum_{\substack{K<l \\ (\neq i,j)}}\sum (1 - Q_K - Q_l) Q_K Q_l \left(\frac{y_K}{Q_K} - \frac{y_i}{Q_i}\right)^2$$

$$< V(t_2) = V_2$$

and likewise,

$$V_n = V(t_n) < V(t_{n-1}) < \cdots < V(t_3) < V(t_2) < V(t_1) = V.$$

So,
$$V(t_D) = \frac{1}{n^2} \sum_{K=1}^{n} V(t_K)$$
$$< \frac{V}{n} = \frac{V(t_{HH})}{n}$$

This result is due to Roychaudhury (1957)

Yet an exact formula for $V(t_D)$ is too difficult to work out but an unbiased estimator for $V(t_D)$ is easy to employ because

$$v(t_D) = \frac{1}{2n^2(n-1)} \sum_{\substack{j \neq j'}}^{n} \sum^{n} (t_j - t_{j'})^2$$

as we already noticed.

One obvious advantage of (PPSWOR, t_D) strategy over (PPSWR, t_{HH}) is that the former involves a sample of only distinct units but in the latter the same unit may appear more than once, a repeated sample adding no new information. Moreover, $V(t_D) < V(t_{HH})$. Also, though it is hard to work out a simple formula for $V(t_D)$, its unbiased estimator is simple in form just as an unbiased estimator for $V(t_{HH})$.

It is of course not easy to compare $V(t_D)$ versus $V_{SRSWOR}(N\bar{y})$ because of a complicated formula for $V(t_D)$ even under a simplistic model we postulated earlier for an illustration. But it is not difficult to unbiasedly estimate the gain in efficiency from a PPSWOR sample taken and surveyed on considering

$$V(t_D) = E(t_d^2) - Y^2 \text{ versus}$$

$$V_{SRSWOR}(N\bar{y}) = N^2 \left(\frac{1}{n} - \frac{1}{N} \right) \sum_{1}^{N} (y_i - \bar{Y})^2 \qquad (1)$$

$$= N^2 \left(\frac{1}{n} - \frac{1}{N} \right) \left[\sum_{1}^{N} y_i^2 - N(\bar{Y})^2 \right]$$

$$= N^2 \left(\frac{1}{n} - \frac{1}{N} \right) \left[\sum_{1}^{N} y_i^2 - \frac{Y^2}{N} \right]. \qquad (2)$$

Y^2 may be unbiasedly estimated by $\delta = t_d^2 - v(t_D)$ and so an unbiased estimator for $V_{SRSWOR}(N\bar{y})$ from a PPSWOR(n) is to be estimated by $N^2 \left(\frac{1}{n} - \frac{1}{N} \right)$
$$\left[\text{an unbiased estimator of } \sum_{1}^{N} y_i^2 \text{ from a PPSWOR}(n) - \frac{\delta}{N} \right].$$

An unbiased estimator for $\sum_{1}^{N} y_i^2$ from a PPSWOR(n) is t_D with every y_i in it replaced by y_i^2.

So, writing this as $t_D\big|_{y_i=y_i^2}$, an unbiased estimator for $V_{SRSWOR}(N\bar{y})$ is

$$\Delta = N^2 \left(\frac{1}{n} - \frac{1}{N}\right)\left[t_D\big|_{y_i=y_i^2} - \frac{\delta}{N}\right].$$

The quantity
$$\text{Diff} = D = \Delta - v(t_D)$$

provides us with a measure of the estimated gain in efficiency in estimating Y by t_D from a PPSWOR over that by estimating Y by $N\bar{y}$ from an SRSWOR.

Both (PPSWOR, t_D) and (PPSWR, t_{HH}) as sampling strategies are handicapped by the fact that t_D is not a function of a sufficient statistic as it is an order static depending on the order in which the units are sampled and t_{HH} uses units that are not all distinct. So, both may be improved upon in the following manner.

Let s be a sample which may involve units which are not necessarily distinct or may involve the knowledge about the order in which its units are selected. Let $t = t(s, \underline{Y})$ be a statistic calculated from the units in such a sample using the respective values of y_i in $\underline{Y} = (y_1, \ldots, y_i, \ldots, y_N)$ but free of the y_i-values for i outside of s.

Let s^* be the set of distinct units in s and the order in which they are selected is just suppressed or ignored. Let $t^* = t^*(s, \underline{Y})$ be a statistic created from such a sample s^* constructed as below given by

$$t^* = t^*(s, \underline{Y}) = t^*(s, \underline{Y})$$
$$= \frac{\sum_{s \to s^*} t(s, \underline{Y})p(s)}{\sum_{s \to s^*} p(s)}$$

for every s to each of which corresponds the same s^* composed of the same units as in s suppressing the order and/or multiplicity with which the units occur in s. By $\sum_{s \to s^*}$ we denote the summation over each of the samples s to each of which corresponds the same s^*. Thus, s^* is derived over amalgamation of several ss that correspond to the same s^* each. By $p(s^*)$, we shall denote $\sum_{s \to s^*} p(s)$.

Now,
$$E(t^*) = \sum_{s^*} p(s^*)t^*(s^*, \underline{Y})$$
$$= \sum_{s^*} \left[\frac{\sum_{s \to s^*} t(s, \underline{Y})p(s)}{\sum_{s \to s^*} p(s)}\right] \sum_{s \to s^*} p(s)$$
$$= \sum_{s^*} \sum_{s \to s^*} t(s, \underline{Y})p(s) = \sum_{s} p(s)t(s, \underline{Y})$$
$$= E(t)$$

Also,
$$E(tt^*) = \sum p(s)t(s, \underline{Y})t^*(s^*, \underline{Y})$$
$$= \sum_{s^*} \sum_{s \to s^*} p(s)t(s, \underline{Y})t^*(s^*, \underline{Y})$$
$$= \sum_{s^*} p(s^*)\left[t^*(s^*, \underline{Y})\right]^2$$
$$= E(t^*)^2$$

So,
$$E(t - t^*)^2 = E(t^2) + E(t^*)^2 - 2E(tt^*)$$
$$= E(t^2) - E(t^*)^2$$

So,
$$V(t) = E(t^2) - (E(t))^2$$
$$= E(t^2) - (E(t^*))^2$$
$$= E(t^*)^2 - (E(t^*))^2 + E(t - t^*)^2$$
$$= V(t^*) + E(t - t^*)^2$$

So, $\qquad V(t^*) \leq V(t)$ because $V(t^*) = V(t) - E(t - t^*)^2$

Unless, $\qquad E(t - t^*)^2 = 0$

ie, $\qquad t = t^*$ with probability equal to 1.

Also, if there exists an unbiased estimator $\nu(t)$ for $V(t)$, i.e.,

if $\qquad\qquad E\nu(t) = V(t),$

then $\qquad\qquad \nu^* = \nu(t) - (t - t^*)^2$

is an unbiased estimator for $V(t^*)$.

If t is t_D, then t_D^* derived from t_D as t^* is from t, then t_D^* is called the symmetrized Des Raj estimator (t_{SDR} in brief).

For $V(t_D)$, an easy unbiased estimator is ν_D and so an easy unbiased estimator for $V(t_D)$ is

$$\nu_{SD}^* = \nu_D - (t_D - t_D^*)^2$$

Of course $\qquad\qquad V(t_{SDR}) \leq V(t_D)$

For a small value of n as 2, 3, 4, it is not difficult to derive t_{SDR} from t_D though it is too hard if n is larger.

Starting with t as the t_{HH} also it is theoretically easy to derive a statistic t^* by the above approach but the procedure is too cumbrous.

It is, of course, easy from a PPSWOR sample to unbiasedly estimate the gain in efficiency of (PPSWOR, t_{SDR}) over (SRSWOR, $N\bar{y}$) as we have shown the corresponding procedure (PPSWOR, t_D) versus (SRSWOR, $N\bar{y}$).

3.5.3 Murthy's Sampling Strategy

Murthy (1957) considered a general sampling scheme for which $p(s)$ is the selection probability of a sample s, $p(s|i)$ is the probability of a sample s of which i is the unit chosen on the first draw with a selection probability p_i, the normed size-measure considered as earlier and $p(s|ij)$ is the selection probability of a sample s of which i and j are units chosen on the first two draws. Then, obviously,

$$p(s) = \sum_{i \in s} p_i p(s|i).$$

For a sample s thus chosen, (Murthy, 1957) gives us for Y the estimator

$$t_M = \frac{1}{p(s)} \sum_{i \in s} y_i p(s|i).$$

This t_M is unbiased for Y because

$$E(t_M) = \sum_s p(s)t_M = \sum_s \sum_{i \in s} y_i p(s|i)$$

$$= \sum_{i=1}^N y_i \left(\sum_{s \ni i} p(s|i) \right) = \sum_{i=1}^N y_i = Y$$

Also, $$V(t_M) = E(t_M - Y)^2$$

$$= \sum_s p(s) \left[\sum_{i=1}^N y_i \left(\frac{p(s|i)}{p(s)} I_{si} - 1 \right) \right]^2$$

writing $$I_{si} = 1 \text{ if } s \ni i$$
$$= 0, \text{ otherwise}$$

So, $$V(t_M) = \sum_{i=1}^N \sum_{j=1}^N y_i y_j d_{ij}$$

writing

$$d_{ij} = \sum_s p(s) \left(\frac{p(s|i) I_{si}}{p(s)} - 1 \right) \left(\frac{p(s|j) I_{sj}}{p(s)} - 1 \right)$$

Let

$$z_i = \frac{y_i}{p_i}. \text{ Then}$$

$$V(t_M) = \sum_i \sum_j d_{ij} p_i p_j z_i z_j$$

If $z_i = C \, \forall i \in U$, i.e., $y_i = C p_i \, \forall i$, then $\sum_{i \in s} p_i p(s|i) = p(s)$ and so $V(t_M) = 0$, i.e.,

$$0 = V(t_M) = C^2 \sum_{i=1}^N \sum_{j=1}^N d_{ij} p_i p_j,$$

i.e., the Non-negative definite quadratic form $\sum_{i=1}^N \sum_{j=1}^N d_{ij} p_i p_j z_i z_j$ subject to

$\sum_{i=1}^N \sum_{j=1}^N d_{ij} p_i p_j = 0$ implies $\sum_{j=1}^N d_{ij} p_i p_j = 0 \forall i$ as may be easily checked. (vide, Chaudhuri (2010), p 73).

As a consequence, $V(t_M)$ may be written as

$$V(t_M) = -\sum_{i<j} \sum d_{ij} p_i p_j (z_i - z_j)^2$$

$$= -\sum_{i<j} \sum d_{ij} p_i p_j \left(\frac{y_i}{p_i} - \frac{y_j}{p_j} \right)^2$$

$$= \sum_{i<j} \sum a_{ij} \left(1 - \sum_{s \ni i,j} \frac{p(s|i) p(s|j)}{p(s)} \right)$$

writing $a_{ij} = p_i p_j \left(\frac{y_i}{p_i} - \frac{y_j}{p_j} \right)^2$.

An unbiased estimator for $V(t_M)$ follows as

$$\nu(t_M) = \sum_{i<j} \sum a_{ij} \frac{I_{s_{ij}}}{p(s)^2} [p(s|ij) p(s) - p(s|i) p(s|j)]$$

because $\sum_{s \ni ij} p(s|ij) = 1 \forall i \neq j \in U$.

3.5.4 A Special Case of Murthy's Strategy: Lahiri's (1951) Ratio, Estimation From his Sample Selection Method

Suppose a unit i from U has been selected by PPS procedure with probability p_i and after selection it is set aside and from the remaining $(N - 1)$ units of U an SRSWOR of $(n - 1)$ be chosen. For this sampling scheme given by Lahiri (1951) though earlier also by Midzuno (1950), one may easily check that

$$p_i = \frac{x_i}{X}, \, p(s|i) = \frac{1}{\binom{N-1}{n-1}}, \, p(s|i, j) = \frac{1}{\binom{N-2}{n-2}}$$

for $j \neq i, i = 1, \ldots, N$ and Murthy's estimator t_M takes the form, on noting $p(s) = \frac{\sum_{i \in s} x_i}{X}$, that

$$t_M = X \frac{\sum_{i \in s} y_i}{\sum_{i \in s} x_i} = X \frac{\bar{y}}{\bar{x}} = t_R, \text{ say.}$$

This t_R is called (Lahiri's, 1951) ratio estimator based on his sampling as above.

For (Lahiri's, 1951) sampling scheme, of course, this Ratio Estimator is exactly unbiased for the population total Y.

Quite interestingly, it follows

from $$V(t_M) = \sum_{i<j}^{N} \sum^{N} p_i p_j \left(1 - \sum_{s \ni i,j} \frac{p(s|i) p(s|j)}{p(s_j)}\right)$$

that $$V(t_R) = \sum_{i<j}^{N} \sum^{N} x_i x_j \left(\frac{y_i}{x_i} - \frac{y_j}{x_j}\right)^2 \left[1 - \sum_{s \ni i,j} \left(\frac{1}{\sum_{i \in s} x_i}\right) \frac{X}{\binom{N-1}{n-1}}\right].$$

An exactly unbiased estimator for this $V(t_R)$ then works out as

$$\nu(t_R) = \sum_{i<j \in s} \sum x_i x_j \left(\frac{y_i}{x_i} - \frac{y_j}{x_j}\right)^2 \left[\binom{N-1}{n-1} - \frac{X}{\sum_{i \in s} x_i}\right] \frac{X}{\sum_{i \in s} x_i}.$$

If N and n both be quite large, calculation of $\nu(t_R)$ will be quite tough.

3.5.5 *General Probability Sampling Schemes and Horvitz–Thompson Estimator*

Let p be a general sampling design admitting positive inclusion probabilities $\pi_i = \sum_{s \ni i} p(s), i \in U$, with

$$\sum_{1}^{N} \pi_i = E(\nu(s)) = \nu \text{ and}$$

$$\pi_{ij} = \sum_{s \ni i,j} p(s), \text{ subject to}$$

$$\sum_{\substack{j=1 \\ j \neq i}}^{N} \pi_{ij} = \sum_{s} p(s)\nu(s)I_{si} - \pi_i$$

$$\sum_{i \neq j} \sum \pi_{ij} = \sum_{s} p(s)\nu^2(s) - \sum_{1}^{N} \pi_i$$

$$= E(\nu^2(s)) - E(\nu(s))$$

$$= V(\nu(s)) + E(\nu(s))[E\nu(s) - 1]$$

$$= V(\nu(s)) + \nu(\nu - 1)$$

$\nu(s) \equiv$ the number of distinct units in a sample s.

In case $\nu(s) = n$ for every s, these reduce, respectively, to

$$\sum_{1}^{N} \pi_i = n, \sum_{\substack{j=1 \\ \neq i}}^{N} \pi_{ij} = (n-1)\pi_i \text{ and}$$

$$\sum_{i \neq j} \sum \pi_{ij} = (n-1).$$

For $Y = \sum_{1}^{N} y_i$, an unbiased estimator given by Horvitz and Thompson (1952) and earlier by Narain (1951) is

$$t_{HT} = \sum_{i \in s} \frac{y_i}{\pi_i} = \sum_{i=1}^{N} \frac{y_i}{\pi_i} I_{si}.$$

Obviously, $E(t_{HT}) = Y$ because $E(I_{si}) = \pi_i$

$$V(t_{HT}) = \sum_1^N \left(\frac{y_i}{\pi_i}\right)^2 V(I_{si})$$

$$+ \sum\sum_{i \neq j} \left(\frac{y_i}{\pi_i}\right)\left(\frac{y_j}{\pi_j}\right) \mathrm{Cov}(I_{si}, I_{sj})$$

$$V(I_{si}) = \pi_i(1 - \pi_i) \text{ because } I_{si}^2 = I_{si}.$$

$$\mathrm{Cov}(I_{si}, I_{sj}) = E(I_{si}I_{sj}) - \pi_i\pi_j = \pi_{ij} - \pi_i\pi_j.$$

$$\text{So, } V(t_{HT}) = \sum_1^N y_i^2 \frac{1 - \pi_i}{\pi_i} + \sum\sum_{i \neq j} \frac{y_i}{\pi_i}\frac{y_j}{\pi_j}(\pi_{ij} - \pi_i\pi_j).$$

This formula is given by Horvitz and Thompson themselves. An exactly unbiased estimator for $V(t_{HT})$ clearly follows as

$$\nu(t_{HT}) = \sum_{i \in s} \frac{y_i^2}{\pi_i}\left(\frac{1 - \pi_i}{\pi_i}\right) + \sum\sum_{i \neq j} \frac{y_i}{\pi_i}\frac{y_j}{\pi_j}\left(\frac{\pi_{ij} - \pi_i\pi_j}{\pi_{ij}}\right)$$

assuming $\pi_{ij} > 0 \forall i \neq j \in U$.

Writing $V(t_{HT})$ again as

$$V(t_{HT}) = \sum_s p(s)\left[\sum_{i=1}^N y_i\frac{I_{si}}{\pi_i} - Y\right]^2$$

$$= \sum_s p(s)\left[\sum_1^N y_i\left(\frac{I_{si}}{\pi_i} - 1\right)\right]^2$$

$$= \sum_1^N\sum_1^N y_iy_j E\left(\frac{I_{si}}{\pi_i} - 1\right)\left(\frac{I_{sj}}{\pi_j} - 1\right)$$

or $\qquad V(t_{HT}) = \sum_i\sum_j d_{ij}\pi_i\pi_jz_iz_j$ on writing

$$z_i = \frac{y_i}{\pi_i} \text{ and } d_{ij} = E\left(\frac{I_{si}}{\pi_i} - 1\right)\left(\frac{I_{sj}}{\pi_j} - 1\right)$$

we may note that if

$$z_i = C\forall i \in U, i.e., y_i = C\pi_i\forall i \in U, \text{ then}$$

$$V(t_{HT}) = 0 \text{ since } \sum_{i \in s} \frac{y_i}{\pi_i} = Cn \text{ and } Y = C\sum_1^N \pi_i.$$

$$\text{So, } V(t_{HT}) = C^2 \sum_i \sum_j d_{ij} \pi_i \pi_j,$$

i.e., the Non-negative definite quadratic form $V(t_{HT}) = \sum_i \sum_j d_{ij} \pi_i \pi_j z_i z_j$ is subject to the condition $\sum_i \sum_j d_{ij} \pi_i \pi_j = 0$. This leads to $\sum_j d_{ij} \pi_i \pi_j = 0 \ \forall i \in U$. This, vide (Chaudhuri, 2010, p. 73), leads to

$$V(t_{HT}) = -\sum_{i<j}^{N} \sum_{}^{N} d_{ij} \pi_i \pi_j \left(\frac{y_i}{\pi_i} - \frac{y_j}{\pi_j} \right)^2$$

$$= \sum_{i<j} \sum \pi_i \pi_j \left(\frac{\pi_i \pi_j - \pi_{ij}}{\pi_i \pi_j} \right) \left(\frac{y_i}{\pi_i} - \frac{y_j}{\pi_j} \right)^2$$

$$= \sum_{i<j} \sum (\pi_i \pi_j - \pi_{ij}) \left(\frac{y_i}{\pi_i} - \frac{y_j}{\pi_j} \right)^2.$$

This form of $V(t_{HT})$ available only for $\nu(s) =$ a constant for every s with $p(s) > 0$ is given by Yates and Grundy (1953). Their unbiased estimator for $V(t_{HT})$ is then

$$\nu_{YG}(t_{HT}) = \sum_{i<j \in s} \sum \left(\frac{\pi_i \pi_j - \pi_{ij}}{\pi_{ij}} \right) \left(\frac{y_i}{\pi_i} - \frac{y_j}{\pi_j} \right)^2$$

provided $\pi_{ij} > 0 \ \forall i \neq j$ in U.

A couple of alternative variance estimators for $V(t_{HT})$ given by Ajgaonkar (1967) are as follows, namely

(1) $\nu_{Aj}(t_{HT}) = \dfrac{1}{\binom{N-2}{n-2}} \dfrac{1}{p(s)} \sum_{i,j \in s} (\pi_i \pi_j - \pi_{ij}) \left(\frac{y_i}{\pi_i} - \frac{y_j}{\pi_j} \right)^2$

when for every s with $p(s) > 0$, $\nu(s) = n$ and

(2)

$$\nu_{Aj}(t_{HT}) = \frac{1}{p(s)} \sum_{i \in s} y_i^2 \left(\frac{1 - \pi_i}{\pi_i} \Big/ \nu_i \right)$$

$$+ \frac{1}{p(s)} \sum_{i \neq j \in s} \sum \frac{y_i \, y_j}{\pi_i \, \pi_j} (\pi_{ij} - \pi_{ij}) \Big/ \nu_{ij}$$

writing $\nu_i =$ Number of samples containing unit i and $\nu_{ij} =$ Number of samples containing both i and $j \, (i \neq j)$ in U.

3.5.6 Rao et al. (1962) Strategy

This is a variety of PPS sampling motivated to ensure (i) each sample to consist of
a common number n of units each distinct, (ii) a handy expression for the variance
of an unbiased estimator based on the Rao-Hartley-Cochran (RHC) scheme for the
population total, (iii) the variance being uniformly smaller than $V(t_{HH})$ based on
PPSWR sampling in n draws and (iv) a uniformly non-negative estimator for the
variance of RHC's estimator of the finite population total.

Rao-Hartley-Cochran (RHC) sampling scheme:
Choose positive integers N_i for $i = 1, \ldots, n$ such that their sum $\sum_n N_i$ equals N.

Take SRSWORs of sizes $N_i, i = 1, \ldots, n$ from the population U of N units without
replacement each. From the ith group of N_i units note the respective units bearing positive integer size-measures x_{ij}s for $j = 1, \ldots, N_i$ and $i = 1, \ldots, n$. Con-
sider $p_{ij} = \frac{x_{ij}}{X_i}$, writing $X_i = \sum_{j=1}^{N_i} x_{ij}$ and write $Q_i = p_{i1} + \cdots + p_{iN_i}$. Then from
the respective ith groups $(i = 1, \ldots, n)$ independently choose just one unit i_j, say,
by PPS method with selection probability $\frac{p_{ij}}{Q_i}$, for $j = 1, \ldots, N_i$ and $i = 1, \ldots, n$.
Then RHC's estimator for Y is

$$t_{RHC} = \sum_n y_{ij} \frac{Q_i}{p_{ij}}$$

writing y_{ij} for the jth $(= 1, \ldots, N_i)$ unit of ith $(= 1, \ldots, n)$ group. Let E_2 denote
conditional expectation for selection of a unit from a fixed group and E_1 the unconditional expectation over choice of the respective n groups.

Then $E_2\left(y_{ij} \frac{Q_i}{p_{ij}}\right) = \sum_{j=1}^{N_i} y_{ij} = Y_i$, say, the sum of the N_i values of the y's for the
units in the ith group.

Then,
$$E_2\left(\sum_n y_{ij} \frac{Q_i}{p_{ij}}\right) = Y_i$$

$$E_1\left[E_2 \sum_n y_{ij} \frac{Q_i}{p_{ij}}\right]$$

$$= E_1\left(\sum_n Y_i\right) = E_1(Y) = Y$$

i.e.,
$$E = E_1 E_2(t_{RHC}) = Y.$$

Thus, t_{RHC} is unbiased for Y.

Now,
$$V(t_{RHC}) = E_1 V_2(t_{RHC}) + V_1 E_2(t_{RHC})$$

$$= E_1 V_2 \left(\sum_n y_{ij} \frac{Q_i}{p_{ij}} \right) + V_1(Y)$$

$$= E_1 \sum_n \left[\sum_{j<k}^{N_i} \sum^{N_i} \frac{p_{ij} p_{ik}}{Q_i Q_i} \left(\frac{y_{ij}}{p_{ij}} Q_i - \frac{y_{ik}}{p_{ik}} Q_i \right)^2 \right]$$

$$= \sum_n \left[\frac{N_i(N_i - 1)}{N(N - 1)} \sum_{i<i'}^{N} \sum^{N} p_i p_{i'} \left(\frac{y_i}{p_i} - \frac{y_{i'}}{p_{i'}} \right)^2 \right]$$

$$= \frac{\sum_n N_i^2 - N}{N(N - 1)} V$$

By Cauchy's and Schwartz's inequality,

$$\left(\sum_n \right) \left(\sum_n N_i^2 \right) \geq N^2$$

$$\text{or} \quad \sum_n N_i^2 \geq \frac{N^2}{n}.$$

$$\text{So, } N_i = \frac{N}{n} \text{ (if } \frac{N}{n} \text{ is an integer)}$$

provides the minimum value of $\sum_n N_i^2$ as $\frac{N^2}{n}$ and hence the minimum value of

$V(t_{RHC}) = \frac{N-n}{(N-1)} \frac{V}{n} < \frac{V}{n} = V(t_{HH})$.

If N is not an integer multiple of n, then the values of N_i will minimize $\sum_n N_i^2$

subject to $\sum_n N_i = N$ for the choice

$$N_i = \left[\frac{N}{n} \right] \text{ for } i = 1, \ldots, K$$

$$= \left[\frac{N}{n} \right] + 1 \text{ for } i = K + 1, \ldots, n.$$

Writing $\left[\frac{N}{n} \right] = \theta$, we get

$$N = \sum_n N_i = K\theta + (n - K)(\theta + 1)$$

or $K = n(\theta + 1) - N$.

So, the minimum value of $\sum_{n} N_i^2$ is

$$K\theta^2 + (n - K)(1 + 2\theta + \theta^2) = n(1 + \theta)^2 - K(1 + 2\theta).$$

So, the minimum value of $\sum_{n} N_i^2 - N$ is $n\theta(1 + \theta) - 2\theta K$.

So, $V(t_{RHC}) = \frac{n\theta(1+\theta)-2\theta K}{[n(1+\theta)-K][n(1+\theta)-K-1]}V$ and this is less than $\frac{V}{n}$.

An unbiased estimator for $V(t_{RHC}) = \frac{\sum_n N_i^2 - N}{N(N-1)}V$ may be derived as follows:

Consider $\qquad a = \sum_{n} Q_i \left(\frac{y_{ij}}{p_{ij}} - t_{RHC} \right)^2$

$$E(a) = E_1 \sum_n \sum_{j=1}^{N_i} Q_i \frac{y_{ij}^2}{p_{ij}^2} \frac{p_{ij}}{Q_i} - E(t_{RHC}^2) \ [\text{ on noting that } \sum_n Q_i = 1]$$

$$= \sum_n \frac{N_i}{N} \sum_1^N \frac{y_i^2}{p_i} - V(t_{RHC}) - Y^2$$

$$= \left(\sum_1^N \frac{y_i^2}{p_i} - Y^2 \right) - V(t_{RHC})$$

$$= V - \frac{\sum_n N_i^2 - N}{N(N-1)}V = \frac{N^2 - \sum_n N_i^2}{N(N-1)}V$$

So, $\qquad \frac{\sum_n N_i^2 - N}{N^2 - \sum_n N_i^2} \sum_n Q_i \left(\frac{y_{ij}}{p_{ij}} - t_{RHC} \right)^2 = v(t_{RHC})$

has $Ev(t_{RHC}) = v(t_{RHC})$.

Thus, $v(t_{RHC})$ is an unbiased estimator for $V(t_{RHC})$.

3.6   Systematic Sampling

(a) Linear,
(b) Circular: with equal and unequal probabilities.

Suppose for the population $U = (1, \ldots, i, \ldots, N)$ of N given units $N = nK$ with both n and K as positive integers greater than or equal to 2. Then in order to take a sample of size n from them, one may typically arrange them as follows:

$$
\begin{array}{ccccccc}
1 & 2 & \cdots & i & \cdots & K \\
K+1 & K+2 & \cdots & K+i & \cdots & 2K \\
\vdots & \vdots & \vdots & \vdots & \vdots & \vdots \\
jK+1 & jK+2 & \cdots & jK+i & \cdots & jK \\
\vdots & \vdots & \vdots & \vdots & \vdots & \vdots \\
\underset{C_1}{(n-1)K+1} & \underset{C_2}{(n-1)K+2} & \cdots & \underset{C_i}{(n-1)K+i} & \cdots & \underset{C_K}{nK}.
\end{array}
$$

We denote by C_i the ith column of the labels $i, K+i, \ldots, (n-1)K+i$ of the $N = nK$ units in U. Each column here consists of n distinct units of U. Choosing at random one of these columns gives a probability sample of n units. Each of these K possible samples taken with a probability $\frac{1}{K}$, i.e., the sample C_i of n units chosen with probability $\mathrm{Prob}(C_i) = \frac{1}{K}$ for $i = 1, \ldots, K$ is called a linear systematic sample with equal probability from U. The sample mean based on the sample C_i is

$$
\bar{y}_i = \frac{1}{n} \sum_{j=0}^{n-1} y_{jK+i}.
$$

Its expectation is

$$
E(\bar{y}_i) = \frac{1}{K} \left[\frac{1}{n} \sum_{i=1}^{K} \sum_{j=0}^{n-1} y_{jK+i} \right]
$$

$$
= \frac{1}{N} \sum_{i=1}^{N} y_i = \bar{Y}.
$$

$$
V(\bar{y}_i) = E[\bar{y}_i - \bar{Y}]^2
$$

$$
= \frac{1}{K} \sum_{i=1}^{K} \left[\frac{1}{n} \sum_{j=0}^{n-1} y_{jK+i} \right]^2 - (\bar{Y})^2
$$

$$
= \frac{1}{K} \sum_{i=1}^{K} \bar{y}_i^{\,2} - (\bar{Y})^2
$$

$$
= \frac{1}{K} \sum_{i=1}^{K} \bar{y}_i^{\,2} - \frac{1}{N^2} \left(\sum_{i=1}^{N} y_i^2 + \sum_{i \neq i'}^{N} \sum^{N} y_i y_{i'} \right).
$$

Since the samples C_i and $C_{i'}$ for every $i \neq i'(= 1, \ldots, K)$ are disjoint no unit of C_i occurring together with any unit of $C_{i'}$, it is impossible for any units of any two distinct samples C_i and $C_{i'} (i \neq i')$ to have a positive inclusion probability in any sample.

Thus, any two units of U occurring in two different samples C_i and $C_{i'}$ $(i \neq i')$ has an inclusion probability zero. So, $\sum\limits_{i \neq j}^{N}\sum^{N} y_i y_j$ cannot be unbiasedly estimated. Hence, $V(\bar{y}_i)$ cannot be unbiasedly estimated from any single linear systematic sample C_i as above.

If N cannot be expressed as $N = nK$ with n and K both as positive integers more than or equal to 2, then alternatively one may take a 'Circular Systematic' equal probability sample as follows:

Take K as a positive integer $\left[\frac{N}{n}\right]$ or as $\left[\frac{N}{n} + 1\right]$. In either case, choose at random one integer R from 1 to N and take your sample from all the n units labelled as

$$a_j = (R + jK) \bmod(N) \text{ for } j = 0, 1, \ldots, (n-1).$$

These labels a_j and N in case a_j turns out zero, constitute a 'Circular Systematic Sample' with equal probability. This is because in this way of sample selection the total number of possible samples is N and each one has the selection probability $\frac{1}{N}$. Whether selected or not, each of the N possible samples of labels may be written down. Then one may count the number of samples n_i that contain the unit i and also may count the number n_{ij} of samples that contain the pair of distinct units (i, j) in the population.

So, for 'Circular Systematic Sample with equal probabilities', $\frac{n_i}{N}$ and $\frac{n_{ij}}{N}$ are, respectively, the inclusion probabilities π_i and π_{ij} for the unit i and pair of units $(i, j), i \neq j$ in $U = (1, \ldots, N)$. An unbiased estimator for $Y = \sum\limits_{i=1}^{N} y_i$ the population total of a variable y then based on a sample s chosen by 'Circular equal probability systematic sampling method' is

$$t_{HT}(\text{SYS}) = \sum_{i \in s} \frac{y_i}{\pi_i} = N \sum_{i \in s} \frac{y_i}{n_i}$$

with a variance

$$V(t_{HT}(\text{SYS})) = \sum_{1}^{N} y_i^2 (\frac{1 - \pi_i}{\pi}) + \sum\sum_{i \neq i'} y_i y_j \left(\frac{\pi_{ij} - \pi_i \pi_j}{\pi_i \pi_j}\right).$$

As in the case of equal probability 'linear systematic sampling' in this 'circular' case also for many pairs of units (i, j), π_{ij} turns out as zero, and hence, no unbiased estimator of the above $V(t_{HT}(\text{SYS}))$ can exist for a single such sample drawn.

If with the respective units of $U = (1, \ldots, i, \ldots, N)$ known positive numbers x_i be associated with i, say, then 'linear' and 'circular' equal probability systematic sampling may be revised as follows as unequal probability 'linear' and 'circular' systematic sampling methods.

Let $X = \sum_{1}^{N} x_i$ and each x_i be revised as a positive integer. Then, if $\frac{X}{n}$ is an integer K, say, then one integer R from 1 through this K may be selected at random.

Let $A_i = \sum_{j=1}^{i} x_j, i = 1, \ldots, K$. Then, consider $a_j = R + jK, j = 0, 1, \ldots,$ $(n-1)$. If $A_{i-1} < a_j \leq A_i$, taking $A_0 = 0$, then the unit i is taken in the sample. This is 'linear' unequal probability systematic sampling, with PPS.

If instead $\frac{X}{n}$ is not an integer, then taking $K = \left[\frac{X}{n}\right]$, one integer R is randomly taken from 1 to X. Then, calculate $a_j = (R + jK) \bmod(X)$. If $a_j = 0$, take unit N in the sample. If $A_{i-1} < a_j \leq A_i$, then unit i is taken in the sample. This is PPS circular systematic sampling.

In linear PPS systematic sampling, the total number of possible samples is $\frac{X}{n}$ and each is selected with an equal probability $\frac{1}{K}$. If n_i, n_{ij} are numbers of realized samples, respectively, containing the unit i and, respectively, the number of samples containing the pair if units (i, j) in U, then

$$\pi_i = \frac{n_i}{K} \text{ and } \pi_{ij} = \frac{n_{ij}}{K}.$$

So, the (Horvitz and Thompson, 1952) estimator for Y along with its variance may be written down. But here also, π_{ij} turns out to have a zero for many pairs (i, j) and so $V(t_{HT})$ cannot have an unbiased estimator from a single linear PPS systematic sample.

For Circular PPS Systematic sampling, X is the total number of possible samples and each is selected with probability $\frac{1}{X}$. So, if n_i and n_{ij} are the numbers of samples, respectively, containing units i and the paired units (i, j) of U, then $\frac{n_i}{X}$ and $\frac{n_{ij}}{X}$ give the inclusion probability π_i for i and that for (i, j). So, the Horvitz-Thompson estimator t_{HT} for Y along with its variance $V(t_{HT})$ can be immediately written down. But π_{ij} turns out zero for many paired units (i, j) and so from a single circular PPS systematic sample $V(t_{HT})$ cannot be unbiasedly estimated.

3.7 Modified Systematic Sampling

To ensure positive inclusion probabilities of paired units

For the systematic sample selection procedures so far described, inclusion probabilities for many paired units (i, j) turn out to be zero. As a consequence, a sample so drawn fails to yield an unbiased variance estimator for an estimated total.

So, taking the clue from Das (1982) and Chaudhuri and Pal (2003a), let us report here a method to modify them so as to remove this shortcoming. For this, let us start with the most general of the procedures, namely 'Circular PPS Systematic Sample' avoiding separate descriptions of the other particular cases which may be covered easily as corollaries.

Let us write $B_i = \sum\limits_{j=1}^{i} x_j(x_{j-1})$ and draw a random number from 1 to $X(X-1)$, of course, taking each x_i as a positive integer.

Let $K = \left[\frac{X(X-1)}{n}\right]$ and calculate $a_j = (R + jK) \bmod (X(X-1))$ for $j = 0, 1, \ldots, n$.

If $B_{i-1} < a_j \le B_i$, then take i into the sample and if a_j equals zero, take the unit N into the sample.

This is 'Modified Circular PPS Systematic Sampling'. Its special cases follow if every x_i is unity and $\frac{X(X-1)}{n}$ is an integer and $\frac{N(N-1)}{n}$ is also an integer. Chaudhuri and Pal (2003) and Das (1982) should be consulted to check that $\pi_{ij} > 0 \, \forall i, j (i \ne j)$ in U for these 'modified' systematic sampling procedures.

For the original and the modified systematic sampling procedures, the sample chosen may not turn out to be composed exclusively of distinct units. So, the realized sample-size may fall short of an intended size n. Sudhakar (1978) first pointed out this possibility. He showed that N and n should be mutually co-primes if an equal probability circular systematic sample may be produced as consisting of each unit distinct from every other.

3.8 Cluster Sampling

Let a finite survey population be supposed to be composed of easily discernible N 'clusters' with the ith cluster consisting of M_i units. Thus, the population size is $M = \sum\limits_{i=1}^{N} M_i$.

Let a sample s of size $M(s) = \sum\limits_{i \in s} M_i$ of units be taken by SRSWOR from the population of M units. Let the sample mean $t_1 = \frac{1}{M(s)} \sum\limits_{i,j \in s} y_{ij}$ of the $M(s)$ sampled units be taken as an unbiased estimator for its expectation, namely

$$E(t_1) = \frac{1}{M} \sum_{i=1}^{N} \sum_{j=1}^{M_i} y_{ij} = \frac{1}{M} \sum_{i=1}^{N} M_i \bar{y}_i = \bar{Y},$$

say, writing $\bar{y}_i = \frac{1}{M_i} \sum\limits_{j=1}^{M_i} y_{ij}$ and observing $\bar{Y}$ above as the population mean.

$$\text{Then, } V(t_1) = \frac{M - M(s)}{M M(s)} \frac{1}{(M-1)} \sum_{i=1}^{N} \sum_{j=1}^{M_i} \left(y_{ij} - \bar{Y}\right)^2. \qquad (3.8.1)$$

Alternatively, let out of the N clusters an SRSWOR of n be chosen and every chosen cluster be entirely surveyed. Using traditionally obvious notations, let

$$t_2 = \frac{N}{M} \frac{1}{n} \sum_{i=1}^{n} \sum_{j=1}^{M_i} y_{ij}$$

$$= \frac{N}{M} \left(\frac{1}{n} \sum_{i=1}^{n} M_i \bar{y}_i \right)$$

be considered as an estimator for $\bar{Y}$.

Then, $E(t_2) = \frac{N}{M} \frac{1}{N} \sum_{i=1}^{N} M_i \bar{y}_i = \bar{Y}$.

Thus, t_2 is also unbiased for $\bar{Y}$.

Also,

$$V(t_2) = \left(\frac{N}{M} \right)^2 \frac{N-n}{Nn} \frac{1}{N-1} \sum_{i=1}^{N} \left(M_i \bar{y}_i - \frac{\sum_{i=1}^{N} M_i \bar{y}_i}{N} \right)^2$$

$$= \left(\frac{N}{M} \right)^2 \left(\frac{1}{n} - \frac{1}{N} \right) \sum_{i=1}^{N} \left(M_i \bar{y}_i - \frac{M}{N} \bar{Y} \right)^2 \Big/ (N-1) \quad (3.8.2)$$

and

$$V(t_1) = \left(\frac{1}{M(s)} - \frac{1}{M} \right) \frac{1}{(M-1)} \sum_{1}^{N} \sum_{j=1}^{M_i} \left(y_{ij} - \bar{Y} \right)^2 \quad (3.8.3)$$

Only in the very unnatural situations when M_i equals $\frac{M}{N}$ for every $i = 1, \ldots, N$ simplifications in the formulae (3.8.2) and (3.8.3) may be noted to offer some ways to analytically compare $V(t_1)$ with $V(t_2)$ in order to reach any suitable conclusions about working out criteria in the construction of the clusters in practice. Chaudhuri (2014) has given suitable algebraic results following classical literature on cluster formations with restrictions to such equal-sized clusters. He reiterated the classical literature to point out that for the sake of efficiency (i) the clusters should be among themselves as unlike of each other as possible contrarily with features of appropriate strata and (ii) the clusters should not be very big in size if a desired level of efficiency is to be retained. A curious reader may consult (Chaudhuri, 2014). But in the present treatise, we are reluctant to allow more spaces to elaborate on this artificial case of equal-sized clusters.

3.9 Multi-stage Sampling

When a finite population is supposed to be composed of a number of 'disjoint parts' together co-extensive with it, the parts are called (1) 'strata' if each of them is sampled independently of each other and (2) 'clusters' or single-stage clusters as a proper sample of these parts is chosen and each chosen part composed of a number of units in it is fully surveyed.

More generally, the units of each selected cluster, regarded as the first-stage units (Fsus) of the population, may be defined as the second-stage units (ssus). If one takes a sample of these ssus already in the chosen fsus, called clusters, then one has two-stage sampling, with no ssu chosen if it is in an fsu which is not selected. If an ssu in its turn is composed of further units out of which samples are chosen calling them the third-stage units (tsus), then we have three-stage sampling. Proceeding similarly adding more stages we have 'Multi-stage sampling' in general. For example, the provinces in a country may be called the clusters or fsus, the districts in each of them may be called the ssus, the villages and cities/towns in each the tsus and the households in each tsu the fourth or the final-stage units (fsus), provided one takes a sample of the fsus or clusters, a sample of the ssus from each sampled fsu (cluster), a sample of tsus from each selected ssu and finally a sample of the Fsus from each selected tsu. This is multi-stage sampling.

The Theory of Multi-stage Sampling

Let $U = (1, \ldots, i, \ldots, N)$ denote a finite population with i as one of its first-stage units (fsus). Let s be a sample of the fsus with $p(s)$ as its selection probability and y_i be the value of a variable y for the ith fsu for $i \in U$. Let E_1, V_1 denote operators for expectation and variance in respect of this first-stage sample selection according to the design p with $p(s)$ as above.

Suppose our problem is to unbiasedly estimate the population total $Y = \sum_1^N y_i$ but the values y_i for $i \in U$ are not ascertainable. But further suppose it is possible to unbiasedly estimate y_i by sampling at later stages 'independently across the selected fsus'. Let E_L, V_L denote operators for expectation and variance in respect of probability sampling at the later stages. Also, let

$$E = E_1 E_L \text{ and } V = E_1 V_L + V_1 E_L$$

be the operators for expectation and variance in respect of probability sampling at the first and the later stages.

Let us consider an estimator $t = \sum_{i \in s} y_i b_{s_i}$ for Y pretending that y_is are ascertainable by dint of sample surveys of the sampled fsus.

Let b_{s_i} be constants free of $\underline{Y} = (y_1, \ldots, y_i, \ldots, y_N)$ but subject to $1 = \sum_{s \ni i} p(s) b_{s_i} \, \forall i \in U$.

Then, $E_1(t) = Y$ and

$$V_1(t) = \sum_{i=1}^{N} y_i^2 C_i + \sum\sum_{i \neq j} y_i y_j d_{ij}$$

writing,

$$C_i = \sum_{s \ni i} p(s) b_{s_i}^2 - 1, i \in U$$

and,

$$d_{ij} = \sum_{s \ni i,j} p(s) b_{s_i} b_{s_j} - 1i, j \in U, i \neq j \in U.$$

Also,

$$\nu(t) = \sum_{i \in s} y_i^2 C_{s_i} + \sum\sum_{i \neq j \in s} y_i y_j d_{s_{ij}}$$

has the property $E_1 \nu(t) = V(t)$ provided one may choose $C_{s_i}, d_{s_{ij}}$ free of $\underline{Y}$ but subject to

$$\sum_{s \ni i} p(s) C_{s_i} = C_i \text{ and } \sum_{s \ni i,j} p(s) d_{s_{ij}} = d_{ij}.$$

Let E_L, V_L be so chosen that by sampling in subsequent stages independently across the selected fsus, is in s such that $\widehat{y}_i$s are available satisfying

$$E_L(\widehat{y}_i) = y_i \text{ and } V_L(\widehat{y}_i) = V_i, i \in s$$

and also are available $\widehat{V}_i, i \in s$ satisfying $E_L(\widehat{V}_i) = V_i$.

Then it is possible to propose for Y the estimator

$$e = \sum_{i \in s} \widehat{y}_i b_{s_i}$$

with the desirable properties

$$Y = E(e) = E_1 E_L(e) = E_1 E_L(\sum_{i \in s} \widehat{y}_i b_{s_i}) = E_1(t)$$

on noting $E_L(e) = t$.

Also,

$$V(e) = E_1 V_L(e) + V_1 E_L(e)$$

$$= E_1(\sum_{i \in s} V_i b_{s_i}^2) + V_1(t)$$

$$= \sum_{1}^{N} V_i C_i + \sum_{1}^{N} V_i + V_1(t)$$

$$= \sum_{1}^{N} (y_i^2 + V_i) C_i + \sum\sum_{i \neq j} y_i y_j d_{ij} + \sum V_i.$$

Then,
$$\nu(e) = \sum_{i \in s}(\widehat{y}_i)^2 C_{s_i} + \sum_{i \neq j \in s}\sum \widehat{y}_i \widehat{y}_j d_{s_{ij}} + \sum_{i \in s} b_{s_i} \widehat{V}_i = \nu_i$$

follows as an unbiased estimator for $V(e)$. To see this, we may note

$$E_L \nu(e) = \sum_{i \in s}(y_i^2 + V_i)C_{s_i} + \sum_{i \neq j}\sum y_i y_j d_{s_{ij}} + \sum_{i \in s} b_{s_i} V_i$$

and

$$E\nu(e) = E_1(E_L\nu(e))$$

$$= \sum_{1}^{N}(y_i^2 + V_i)C_i + \sum_{i \neq j}\sum y_i y_j d_{ij} + \sum V_i.$$

It is instructive to start with the (Horvitz and Thompson, 1952) estimator for the sampling of fsus

$$t_{HT} = \sum_{i \in s}\frac{y_i}{\pi_i}$$

for a sampling design p with $\pi_i > 0 \forall i$ and $\pi_{ij} > 0 \forall i, j (i \neq j) \in U$ pretending that y_is are known or available on surveying the fsus sampled.

Then, $E_1(t_{HT}) = Y$ and

$$V_1(t_{HT}) = \sum_{1}^{N} y_i^2 \left(\frac{1-\pi_i}{\pi_i}\right) + \sum_{i \neq j}\sum y_i y_j \left(\frac{\pi_{ij} - \pi_i \pi_j}{\pi_i \pi_j}\right).$$

But since y_is are really unavailable, suppose sampling is carried out in subsequent stages deriving unbiased estimator $\widehat{y}_i$ for y_i independently across the selected fsus i in a first sample s such that $E_L(\widehat{y}_i) = y_i$, for i in s, $V_L(\widehat{y}_i) = V_i$ and estimates ν_i for i in s are available such that $E_L(\nu_i) = V_i$ for i in s, then one may employ for Y the unbiased estimator

$$e = \sum_{i \in s}\frac{\widehat{y}_i}{\pi_i}$$

because $E_L(e) = t_{HT}$ and
$$E(e) = E_1(E_L(e)) = E_1(t_{HT}) = Y.$$

Now, $V_L(e) = \sum_{i \in s}\frac{V_i}{\pi_i^2}$ and

$$V(e) = V_1 E_L(e) + E_1 V_L(e)$$

$$= V_1(t_{HT}) + E_1 \left(\sum_{i \in s} \frac{V_i}{\pi_i^2} \right)$$

$$= \sum_1^N y_i^2 \left(\frac{1 - \pi_i}{\pi_i} \right) + \sum_{i \neq j \in s}^N \sum^N y_i y_j \left(\frac{\pi_{ij} - \pi_i \pi_j}{\pi_i \pi_j} \right) + \sum_1^N \frac{V_i}{\pi_i}$$

$$\nu(e) = \sum_{i \in s} (\widehat{y_i})^2 \left(\frac{1 - \pi_i}{\pi_i} \right) \frac{1}{\pi_i}$$

$$+ \sum_{i \neq j \in s}^N \sum^N \widehat{y_i} \widehat{y_j} \left(\frac{\pi_{ij} - \pi_i \pi_j}{\pi_i \pi_j} \right) + \frac{1}{\pi_{ij}} + \sum_{i \in s} \frac{\nu_i}{\pi_i}$$

is an unbiased estimator for $V(e)$.

This is because

$$E_L \nu(e) = \sum_{i \in s} y_i^2 \left(\frac{1 - \pi_i}{\pi_i} \right) \frac{1}{\pi_i} + \sum_{i \in s} V_i \left(\frac{1 - \pi_i}{\pi_i} \right) \frac{1}{\pi_i}$$

$$+ \sum_{i \neq j \in s} \sum \frac{y_i y_j}{\pi_i \pi_j} \left(\frac{\pi_{ij} - \pi_i \pi_j}{\pi_i \pi_j} \right) + \sum_{i \in s} \frac{V_i}{\pi_i} \text{ and}$$

$$E \nu(e) = E_1(E_L \nu(e))$$

$$= \sum_1^N y_i^2 \left(\frac{1 - \pi_i}{\pi_i} \right) + \sum_1^N V_i \left(\frac{1 - \pi_i}{\pi_i} \right)$$

$$+ \sum_{i \neq j}^N \sum^N y_i y_j \left(\frac{\pi_{ij} - \pi_i \pi_j}{\pi_i \pi_j} \right) + \sum_1^N V_i$$

$$= V(e).$$

3.10 Ratio and Regression Methods of Estimation

In the sample survey, we come across variables that are directly or indirectly correlated with the study variables. These variables are called auxiliary variables. Using
the information on auxiliary variables, we may improve the estimation of the total
or mean of the study variable y. The auxiliary variable is denoted by x here.

For example, suppose it is desired to estimate the total yield of paddy in a certain
region. The information on the total cultivated area is also available. Denoting Y and
X as total yield of paddy and total cultivated area, we may define the ratio as $R = \frac{Y}{X}$
where X is known. The sampling design may be simple random or complex.

In estimating ratios, the commonly used procedure is to take the ratios of unbiased
estimators of numerator and denominator. It is termed as ratio estimator denoted as
$\widehat{R} = \frac{\widehat{Y}}{\widehat{X}}$, where $\widehat{Y}$ and $\widehat{X}$ are unbiased estimators of Y and X, respectively.

The ratio estimator of the population total Y may be taken as $\widehat{Y}_R = \widehat{R}X$ as X is known.

To judge the estimator $\widehat{R}$ analytically, we will define e and e' as $e = \frac{\widehat{Y}-Y}{Y}$ and $e' = \frac{\widehat{X}-X}{X}$.

Then the estimator $\widehat{R}$ nay be written as $\widehat{R} = \frac{Y(1+e)}{X(1+e')} = R(1+e)(1+e')^{-1}$.

Expanding $(1+e')^{-1}$, we may write $\widehat{R}$ as

$$\widehat{R} = R(1+e)(1-e'+e'^2) \text{ taking } |e'| < 1.$$

So
$$E(\widehat{R}) = R[E(e')^2 - E(ee') + 1] \text{ as } E(e) = E(e') = 0$$
$$= R[\frac{V(\widehat{X})}{X^2} - \text{Cov}(\widehat{X}, \widehat{Y}) + 1].$$

So that the bias of $\widehat{R}$ is

$$B(\widehat{R}) = E(\widehat{R} - R)$$
$$= R\left[\frac{V(\widehat{X})}{X^2} - \frac{\text{Cov}(\widehat{X}, \widehat{Y})}{XY}\right]$$
$$= \frac{1}{X^2}\left[RV(\widehat{X}) - \text{Cov}(\widehat{X}, \widehat{Y})\right].$$

The terms involving second and higher powers of (e, e') in the above expression would be negligible due to the fact that e and e' are likely to be small quantities when the sample-size is large.

Also, the Mean Square Error (MSE) of $\widehat{R}$ is

$$M(\widehat{R}) = E(\widehat{R} - R)^2 = R^2 E[(1+e)(1+e')^{-1}]$$
$$= R^2 E(e - e')^2$$
$$= R^2 E(e^2 - 2ee' + e'^2)$$
$$= \frac{1}{X^2}\left[V(\widehat{Y}) - 2R\,\text{Cov}(\widehat{X}, \widehat{Y}) + R^2 V(\widehat{X})\right].$$

Neglecting the terms of degree greater than 2 in (e, e') and assuming the bias negligibly small, the MSE reduces to variance of R.

The term $M(\widehat{R})$ may be estimated as

$$\widehat{M}(\widehat{R}) = \frac{1}{\widehat{X}^2}[\widehat{V}(\widehat{Y}) - 2\widehat{R}\widehat{\text{Cov}}(\widehat{X}, \widehat{Y}) + \widehat{R}^2\widehat{V}(\widehat{X})]$$

where unbiased estimators of $V(\widehat{Y})$, $V(\widehat{X})$ and $\text{Cov}(\widehat{X}, \widehat{Y})$ are available.

Method of interpenetrating sub-sample may be also used to get unbiased estimator of $M(\widehat{R})$.

Let the initial sample s be union of m independent samples as drawn which are denoted as $s_1, s_2, \ldots, s_m$. Let $\widehat{Y}_1, \widehat{Y}_2, \ldots, \widehat{Y}_m$ and $\widehat{X}_1, \widehat{X}_2, \ldots, \widehat{X}_m$ be m unbiased estimators of the totals Y and X, based on m samples above.

Then $\widehat{X} = \frac{1}{m} \sum_1^m \widehat{X}_i$ and $\widehat{Y} = \frac{1}{m} \sum_1^m \widehat{Y}_i$ become unbiased estimators of X and Y, respectively.

By method of interpenetrating sub-sampling, $V(\widehat{Y})$ may be estimated as $\widehat{V}(\widehat{Y}) = \frac{1}{m(m-1)} \sum_1^m (\widehat{Y}_i - \widehat{Y})^2$.

Then finally $\widehat{V}(\widehat{R})$ is written as

$$\widehat{V}(\widehat{R}) = \widehat{V}\left(\frac{\widehat{Y}}{\widehat{X}}\right) = \frac{1}{\widehat{X}^2} \widehat{V}(\widehat{Y})$$

$$= \frac{1}{\widehat{X}^2} \frac{1}{m(m-1)} \sum_1^m (\widehat{Y}_i - \widehat{Y})^2.$$

Alternatively, defining $\widehat{R}_i = \frac{\widehat{Y}_i}{\widehat{X}_i}, 1, 2, \ldots, m$, an alternative ratio estimator based on m interpenetrating sub-sampling may be termed as

$$\widehat{R}' = \frac{1}{m} \sum_{i=1}^m \widehat{R}_i.$$

The related variance estimator is

$$\widehat{V}(\widehat{R}') = \frac{1}{m(m-1)} \sum_1^m (\widehat{R}_i - \widehat{R}')^2.$$

Taking $m = 2$, the estimator is called the estimator based on half sampling. Indian National Sample Survey Organization (NSSO) follows the 'half sampling' for simplicity. This method may be applied to any complex survey.

Theorem *The ratio estimator $\widehat{Y}_R$ of the population total is more efficient than $\widehat{Y}$ if the correlation coefficient between $\widehat{X}$ and $\widehat{Y}$, say, $P(\widehat{X}, \widehat{Y}) < \frac{-1}{2} \frac{C(\widehat{X})}{C(\widehat{Y})}$ where $C(\widehat{X})$ and $C(\widehat{Y})$ are relative standard errors of $\widehat{X}$ and $\widehat{Y}$, respectively and $P(\widehat{X}, \widehat{Y})$ is negative.*

Proof The MSE of $\widehat{Y}_R$ is

$$M(\widehat{Y}_R) = M(\widehat{R}X) = X^2 M(\widehat{R})$$
$$= V(\widehat{Y}) - 2R\mathrm{Cov}(\widehat{X}, \widehat{Y}) + R^2 V(\widehat{X}).$$

$\widehat{Y}_R$ will be more efficient than $\widehat{Y}$.

Case 1: If R is positive, $\widehat{Y}_R$ is more efficient than $\widehat{Y}$ if

$$2R\mathrm{Cov}(\widehat{X}, \widehat{Y}) > R^2 V(\widehat{X})$$

or

$$\frac{\mathrm{Cov}(\widehat{X}, \widehat{Y})}{V(\widehat{X})} > \frac{R}{2}$$

or

$$P(\widehat{X}, \widehat{Y}) > \frac{1}{2}\frac{C(\widehat{X})}{C(\widehat{Y})}$$

where

$$C(\widehat{X}) = \frac{\mathrm{s.e}(\widehat{X})}{\widehat{X}}.$$

and similarly $C(\widehat{Y})$.

Case 2: If R is negative, $\widehat{Y}_R$ is more efficient than $\widehat{Y}$ if

$$P(\widehat{X}, \widehat{Y}) < \frac{-1}{2}\frac{C(\widehat{X})}{C(\widehat{Y})}.$$

In case of stratified sampling, the estimator $\widehat{R}$ is defined as

$$\widehat{R} = \frac{\widehat{Y}}{\widehat{X}} = \frac{\sum_1^L \widehat{Y}_h}{\sum_1^L \widehat{X}_h}$$

assuming L strata in which $\widehat{X}_h$ and $\widehat{Y}_h$ are the unbiased estimators of the hth stratum total X_h and Y_h, respectively, $h = 1, \ldots, L$.

The ratio estimator under stratified random sampling may be termed as

$$\widehat{Y}_{st(R)} = \widehat{R}X = \left(\frac{\sum_1^L \widehat{Y}_h}{\sum_1^L \widehat{X}_h}\right).X.$$

Under stratified sampling $V(\widehat{Y}) = \sum_{h=1}^{L} V(\widehat{Y}_h)$, $V(\widehat{X}) = \sum_1^L V(\widehat{X}_h)$ and $\mathrm{Cov}(\widehat{X}, \widehat{Y}) = \sum_1^L \mathrm{Cov}(\widehat{X}_h, \widehat{Y}_h)$.

The bias and the variance of $\widehat{Y}_R$ are

$$B(\widehat{Y}_R) = \frac{1}{X^2}\sum_{h=1}^{L}\left[RV(\widehat{X}_h) - \mathrm{Cov}(\widehat{X}_h, \widehat{Y}_h)\right]$$

and

$$V(\widehat{Y}_R) = \sum_{h=1}^{L}\left[V(\widehat{Y}_h) - 2R\mathrm{Cov}(\widehat{X}_h, \widehat{Y}_h) + R^2 V(\widehat{X}_h)\right]$$

where $R = \frac{\sum_{h=1}^{L} Y_h}{\sum_{h=1}^{L} X_h}$.

The above method of estimation is called Combined Ratio estimator based on pooled or combined ratio estimation of $\widehat{Y}$ and $\widehat{X}$.

An alternative separate ratio estimator may be defined in which estimators are separately built up for stratum level ratios.

The separate ratio estimator may be defined as

$$\widehat{Y}_{R'} = \sum_{h=1}^{L} \widehat{R}_h X_h = \sum_{h=1}^{L} \frac{\widehat{Y}_h}{\widehat{X}_h} . X_h .$$

The bias and variances are

$$B(\widehat{Y}_{R'}) = \sum_{1}^{L} \frac{1}{X_h} \left[R_h V(\widehat{X}_h) - \text{Cov}(\widehat{X}_h, \widehat{Y}_h) \right]$$

and

$$V(\widehat{Y}_{R'}) = \sum_{h=1}^{L} \left[V(\widehat{Y}_h) - 2R_h \text{Cov}(\widehat{Y}_h, \widehat{X}_h) + R_h^2 V(\widehat{X}_h) \right]$$

where $R_h = \frac{Y_h}{X_h}$ and $\widehat{R}_h = \frac{\widehat{Y}_h}{\widehat{X}_h}$ for hth stratum $h = 1, 2, \ldots, L$.

N.B: Under simple random sampling with replacement (SRSWR) scheme (of sample-size n), the ratio estimator $(\widehat{R})$ may be written as

$$\widehat{R} = \frac{\widehat{y}}{\widehat{x}},$$

defining $(\widehat{y})$ and $(\widehat{x})$ as the sample-based estimators of the population means $(\bar{y})$ and $(\bar{x})$, respectively.

The variance $V(\widehat{R})$ may be written as

$$V(\widehat{R}) = \frac{1}{\bar{x}^2} \left(\frac{\sigma y^2}{n} - 2RP \frac{\sigma x \sigma y}{n} + R^2 \frac{\sigma x^2}{n} \right)$$

defining σy^2, σx^2 as $\sigma y^2 = \frac{1}{N} \sum_{i=1}^{N} (y_i - \bar{Y})^2$ and $\sigma x^2 = \frac{1}{N} \sum_{i=1}^{N} (x_i - \bar{X})^2$. P is referred to as population correlation coefficient between x and y.

The bias term is written as

$$B(\widehat{R}) = \frac{1}{\bar{x}^2 n} (R\sigma x^2 - \sigma xy)$$

where $\sigma xy = \frac{1}{N(N-1)} \sum_{i=1}^{N} (y_i - \bar{Y})(x_i - \bar{X})$.

Under the above SRSWR scheme, the terms $V(\widehat{R})$ and $B(\widehat{R})$ are estimated by

$$\widehat{V}(\widehat{R}) = \frac{1}{\bar{x}^2}\frac{1}{n}(sy^2 - 2\widehat{R}sxy + \widehat{R}^2sx^2)$$

$$= \frac{1}{\bar{x}^2}\sum_{i=1}^{n}\frac{(y_i - \widehat{R}x_i)^2}{n(n-1)}$$

and

$$\widehat{B}(\widehat{R}) = \frac{1}{\bar{x}^2}\frac{1}{n}(\widehat{R}sx^2 - sxy)$$

$$= \frac{1}{\bar{x}^2}\sum_{i=1}^{n}\frac{x_i(\widehat{R}x_i - y_i)}{n(n-1)}.$$

Similarly, under varying probability sampling scheme (say, probability proportionate to size with replacement scheme, PPSWR), the ratio estimator $\widehat{R}$ may be modified as

$$\widehat{R} = \frac{\sum_{i=1}^{n}\frac{y_i}{p_i}}{\sum_{i=1}^{n}\frac{x_i}{p_i}}$$

defining p_i as the normed size-measure under PPSWR scheme.

The variance $V(\widehat{R})$ and bias $B(\widehat{R})$ may be modified as

$$V(\widehat{R}) = \frac{1}{n\bar{x}^2}\sum_{i=1}^{N}\frac{1}{p_i}(y_i - Rx_i)^2$$

and

$$B(\widehat{R}) = \frac{1}{n\bar{x}^2}\sum_{i=1}^{N}\frac{x_i}{p_i}(Rx_i - y_i).$$

The ratio estimator under the PPSWR scheme gives us efficient results in estimation if the set of probabilities used for selection of PPS sample is appropriate and if $\frac{y_i}{p_i}$ and $\frac{x_i}{p_i}$ are positively correlated for ith unit.

Regression Estimators: Using the information on an auxiliary variable x that is correlated with y, the method of estimation may be improved by a regression estimator. When the variable y is linearly related with the auxiliary variable x but the line does not go through the origin, then the estimate based on the linear regression of y on x is recommended.

An unbiased estimator for the population total Y may be suggested as

$$\widehat{Y}' = \widehat{Y} + \lambda(\widehat{X} - X)$$

where λ is a constant

$$X = \sum_{i=1}^{N}x_i,$$

and $\widehat{Y}, \widehat{X}$ are related unbiased estimators of the population totals Y and X, respectively.

The variance of $\widehat{Y}'$ is

$$V(\widehat{Y}') = V(\widehat{Y}) + 2\lambda \text{Cov}(\widehat{X}, \widehat{Y}) + \lambda^2 V(\widehat{X}).$$

λ can be determined by minimizing $V(\widehat{Y}')$.

Here writing $\frac{\delta V(\widehat{Y})}{\delta \lambda} = 0$ gives us

$$\widehat{\lambda} = \frac{-\text{Cov}(\widehat{X}, \widehat{Y})}{V(\widehat{X})} = -\beta,$$

where β is the regression coefficient of $\widehat{Y}$ on $\widehat{X}$.

So the regression estimator $\widehat{Y}'$ may be deduced as

$$\widehat{Y}'_r = \widehat{Y} + \widehat{\beta}(X - \widehat{X})$$

where $\widehat{\beta}$ is the defined by

$$\widehat{\beta} = \frac{\widehat{\text{Cov}(\widehat{X}, \widehat{Y})}}{\widehat{V}(\widehat{X})} = \rho \frac{\widehat{\sigma}_Y}{\widehat{\sigma}_X}.$$

Here $\widehat{\text{Cov}}(\widehat{X}, \widehat{Y})$ and $\widehat{V}(\widehat{X})$ are the estimates of $\text{Cov}(\widehat{X}, \widehat{Y})$ and $V(\widehat{X})$, respectively.

Defining $e = \frac{\widehat{Y}-Y}{Y}$, $e' = \frac{\widehat{X}-X}{X}$ and $e'' = \frac{\widehat{\beta}-\beta}{\beta}$, we can write the regression estimator $\widehat{Y}'_r$ as

$$\begin{aligned} \widehat{Y}'_r &= \widehat{Y} + \widehat{\beta}(X - \widehat{X}) \\ &= Y(1 + e) + \widehat{\beta}(1 + e'')[X - X(1 + e')] \\ &= Y(1 + e) + \widehat{\beta}(1 + e'')X(-e'). \end{aligned}$$

So $E(\widehat{Y}'_r) \simeq Y$ as $E(e) = E(e') = E(e'') = 0$ and the variance of $\widehat{Y}'_r$ is

$$\begin{aligned} V(\widehat{Y}'_r) &= E(\widehat{Y}'_r - Y)^2 \\ &= V(\widehat{Y}) + \beta^2 V(\widehat{X}) - 2\text{Cov}(\widehat{X}, \widehat{Y}) \\ &= (1 - \rho^2)V(\widehat{Y}) \text{ writing } \beta = \rho \frac{\sigma_Y}{\sigma_X}. \end{aligned}$$

Under SRSWOR sampling scheme, the regression estimator $\widehat{Y}_r'$ can be written as

$$\widehat{Y}_r = N\bar{y} + N\widehat{\beta}(\bar{X} - \bar{x})$$

where
$$\widehat{\beta} = \frac{\sum_{i=1}^{n}(x_i - \bar{x})(y_i - \bar{y})}{\sum_{i=1}^{n}(x_i - \bar{x})^2}$$

and
$$V(\widehat{Y}_r) = N^2(1 - f)\frac{Sy^2(1 - \rho^2)}{n}$$

where f is the finite population correction (f.p.c) which is equal to $\frac{n}{N}$.

The bias term is
$$B(\widehat{Y}_r) = -\text{Cov}(\widehat{\beta}, x).$$

Here $V(\widehat{Y}_r)$ nay be unbiasedly estimated as

$$V(\widehat{Y}_r) = \frac{N^2(1 - f)}{n(n - 2)}\sum_{i=1}^{n}\left\{(y_i - \bar{y}) - \widehat{\beta}(x_i - \bar{x})\right\}^2.$$

Under stratified random sampling, the combined regression estimator is

$$\widehat{Y}_r = \sum_{h=1}^{L} N_h \bar{y}_h + \widehat{\beta}\left(X - \sum_{h=1}^{L} N_h \bar{x}_h\right)$$

where
$$\widehat{\beta} = \frac{\sum_{h=1}^{L}\widehat{W}_h\left(\sum_{i=1}^{n_h}(y_{hi} - \bar{y}_h)(x_{hi} - \bar{x}_h)\right)}{\sum_{h=1}^{L}\widehat{W}_h\sum_{i=1}^{n_h}(x_{hi} - \bar{x}_h)^2}$$

and
$$\widehat{W}_h = \frac{N_h^2(1 - f_h)}{n_h(n_h - 1)}, f_h = \frac{n_h}{N_h}$$

The $V(\widehat{Y}_r)$ is

$$V(\widehat{Y}_r) = \sum_{h=1}^{L} N_h^2(1 - f_h)\frac{S_h^2(1 - \rho^2)}{n_h}.$$

The separate regression estimator is

$$\widehat{Y}_r^* = \sum_{h=1}^{L} N_h\{\bar{y}_h) - \widehat{\beta}_h(\bar{X}_h - \bar{x}_h)\}$$

and
$$V(\widehat{Y}_r^*) = \sum_{h=1}^{L} N_h^2(1 - f_h)(1 - \rho_h^2)\frac{S_h^2}{n_h}$$

where ρ_h is the correlation coefficient between y and x in the h_{th} statrum.

Comparison:
$$V(\widehat{Y}_R) - V(\widehat{Y}_r') = (\rho\sigma(\widehat{Y}) - R\sigma(\widehat{X}))^2$$

The regression estimator performs better than ratio estimator as the above term is always positive.

Equality occurs if $\beta = R$, i.e., the line of regression passes through the origin.

3.11 Almost Unbiased Ratio Estimators

The bias of a ratio estimator may be unbiasedly estimated (to the second order of approximation) by the technique of replicated (or interpenetrating) samples, introduced by Mahalanobis (1939). The technique consists of drawing two or more samples from the population. Suppose the sample is drawn in the form of m independent interpenetrating sub-samples of the same size and with the same design. Let $t_1, t_2, \ldots, t_m$ be the uncorrelated random variables with the same expectation $E(t_i) = \theta \, \forall i = 1, 2, \ldots, m$.

Let $\bar{t} = \frac{1}{m} \sum_{i=1}^{m} t_i$. Then an unbiased estimator of $V(\bar{t})$ is given by $\widehat{V}(\bar{t}) = \frac{1}{m(m-1)} \sum_{i=1}^{m} (t_i - \bar{t})^2$. Later (Lahiri, 1954; Koop, 1960; Deming, 1956) and others discussed its modified versions in several survey sampling contexts.

In ratio estimation context, the estimator of the bias may be used to correct the ratio estimator for its bias. A modified ratio estimator may be derived which is unbiased upto the second order of approximation. Let $\{\widehat{Y}_i\}$ and $\{\widehat{X}_i\}(1, 2, \ldots, m)$ be the unbiased estimates of the population totals Y and X based on m replicated sub-samples. Two estimators R_1 and R_m may be used to estimate the ratio $R = \frac{Y}{X}$, where $\widehat{R}_1 = \frac{\widehat{Y}}{\widehat{X}}$ and $\widehat{R}_m = \frac{1}{m} \sum_{i=1}^{m} \frac{\widehat{Y}_i}{\widehat{X}_i}$ defining $\widehat{Y} = \frac{1}{m} \sum_{i=1}^{m} \widehat{Y}_i$ and $\widehat{X} = \frac{1}{m} \sum_{i=1}^{m} \widehat{X}_i$.

$\widehat{R}_m$ may be written as $\widehat{R}_m = \frac{1}{m} \sum_{i=1}^{m} r_i$, where $r_i = \frac{\widehat{Y}_i}{\widehat{X}_i}$.

The bias term of $\widehat{R}_1$ may be written as

$$
\begin{aligned}
B_1 = B(\widehat{R}_1) &= \frac{1}{X^2}[RV(\widehat{X}) - \text{Cov}(\widehat{X}, \widehat{Y})] \\
&= \frac{1}{m^2} \frac{1}{X^2} \sum_{i=1}^{m} \{RV(\widehat{X}_i) - \text{Cov}(\widehat{X}_i, \widehat{Y}_i)\} \\
&= \frac{1}{m^2} \sum_{i=1}^{m} B(r_i) \\
&= \frac{1}{m} \left(\frac{1}{m} \sum_{i=1}^{m} B(r_i) \right) \\
&= \frac{1}{m} B_m
\end{aligned}
$$

where $\qquad B_m = \text{Bias of } \widehat{R}_m$

$$= \frac{1}{m} \sum_{i=1}^{m} B(r_i).$$

We may write $m B_1 = B_m$ from above

$$E(\widehat{R}_m - \widehat{R}_1) = B_m - B_1$$
$$= (m - 1)B_1.$$

The term B_1 may be unbiasedly estimated by

$$\widehat{B}_1 = \frac{1}{m - 1}(\widehat{R}_m - \widehat{R}_1).$$

So the ratio estimator $\widehat{R}_1$ may be corrected for its bias. The corrected estimator $\widehat{R}_C$ is

$$\widehat{R}_C = \widehat{R}_1 - \widehat{B}_1 = \frac{1}{m - 1}(m\widehat{R}_1 - \widehat{R}_m).$$

$\widehat{R}_C$ may be used as unbiased estimator and it is known as almost unbiased ratio estimator. The term 'almost' is used as it is unbiased only second order of approximation.

The variance terms of $\widehat{R}_1$ and $\widehat{R}_m$ are as follows:

$$V(\widehat{R}_1) = \frac{1}{m^2 X^2} \sum_{i=1}^{m} \{V(\widehat{Y}_i) - 2\mathrm{Cov}(\widehat{X}_i, \widehat{Y}_i) + R^2 V(X_i)\}$$

$$= \frac{1}{m^2} \sum_{i=1}^{m} V(r_i)$$

$$= V(\widehat{R}_m).$$

So the variances of $\widehat{R}_1$ and $\widehat{R}_m$ correct to their first order of approximation are the same. But the estimator $\widehat{R}_1$ is preferred as it has less bias than the other.

3.12   Multi-phase Sampling

Two-phase or double sampling was introduced by Neyman (1938). Multi-phase or multiple sampling is an obvious extension to the same.

To tackle some aspects of stratified sampling, ratio estimation, regression estimation and an often occurring phenomenon of failing to observe data in surveying an initially chosen sample at least in certain parts of it an effective procedure is to resort to double sampling. Let us narrate the related possibilities.

3.12.1 Double Sampling in Stratification

Suppose a finite population of N units $U = (1, \ldots, i, \ldots, N)$ is conceptually composed of H strata, the hth stratum ($h = 1, 2, \ldots, H$) consisting of N_h units ($\sum_{h=1}^{H} N_h = N$) composed of those with y-values such that $a_{h-1} < y_i \leq a_h$, for i in U and a_h-values pre-specified, $a_0 > -\infty$ and $a_0 < a_1 \cdots < a_H < +\infty$, $h = 1, \ldots, H$. Suppose $W_h = \frac{N_h}{N}, h = 1, \ldots, H$ but neither $N_h(h = 1, \ldots, H)$ nor the y_i-values are known. If we know $W_h(h = 1, \ldots, H)$ and could identify the units in the respective strata $(a_{h-1}, \; a_h]$ for $h = 1, \ldots, H$, then we could employ a standard stratified SRSWOR with n_h units sampled from the h_{th} stratum and adopt for $\bar{Y} = \frac{\sum_1 y_i}{N} = \frac{Y}{N}$ the standard unbiased estimator

$$\bar{y}_{st} = \frac{1}{N} \sum_1^H N_h \bar{y}_h = \sum_1^H W_h \bar{y}_h$$

recalling $\bar{Y} = \sum_1^H W_h \bar{Y}_h$, $\bar{y}_h$, $\bar{Y}_h$ the sample and population means for the h_{th} stratum, $h = 1, \ldots, H$.

Alternatively, therefore, we may estimate the unknown W_hs and use these estimates in deriving a suitable estimator for Y in the following manner:

Take a first-hand SRSWOR of n' units from the population and surveying it find the numbers $n'_h(h = 1, \ldots, H)$ of them having appeared from the respective strata $h(= 1, \ldots, H)$. Then $W'_h = \frac{n'_h}{n'}$, for $h = 1, \ldots, H$ have the generalized Hypergeometric distribution HG $(n', W_h, h = 1, \ldots, H)$ so that $E_1(W'_h) = W_h$, writing E_1 as the expectation operator for the first-phase sampling; later by V_1 we shall denote the corresponding variance operator.

In the second phase, let from the respective n'_h units found from the strata $h = 1, \ldots, H$ based on the first-phase sample independently across the strata SRSWORs of n_h units be selected and the sample means of ys based on these n_h second-phase samples-based units, namely $\bar{y}_h, h = 1, \ldots, H$ be obtained.

Then $\bar{y}_{dst} = \sum_{h=1}^{H} W'_h \bar{y}_h$ may be taken as an unbiased estimator for $\bar{Y}$. This is because writing E_2, V_2 as expectation and variance operators for this sampling in the second phase and $E = E_1 E_2, V = E_1 V_2 + V_1 E_2$ as the expectation and variance for the above Two-phase or the Double sampling thus executed, we have $E_2(\bar{y}_h) = \bar{y}'_h =$ the mean of all the ys in the first-phase sample of n'_h units observed from the h_{th} stratum

$$E_2(\bar{y}_{dst}) = \sum_{h=1}^{H} W'_h \bar{y}'_h = \frac{1}{n'} \sum_{i=1}^{n'} y_i = \bar{y}', \text{ say}$$

$$E(\bar{y}_{dst}) = E_1(\bar{y}') = \frac{1}{N} \sum_{i=1}^{N} y_i = \bar{Y},$$

$$V(\bar{y}_{dst}) = V_1(\bar{y}') + E_1[V_2(\bar{y}_{dst})]$$

$$= \frac{N - n'}{Nn'} S^2$$

$$+ E_1 \left[\sum_1^H \left(\frac{n'_h}{n'}\right)^2 \frac{\left(\frac{1}{n_h} - \frac{1}{n'_h}\right)}{(n'_h - 1)} \sum_1^H \left(y_{h_i} - \bar{y}'_h\right)^2 \right]$$

writing
$$S^2 = \frac{1}{N - 1} \sum_{i=}^{N} (y_i - \bar{Y})^2$$

$$= \frac{1}{N - 1} \sum_{h=1}^{H} \sum_{i=}^{N_h} (y_{h_i} - \bar{Y})^2,$$

$$y_i \ = \ \text{value of } y \text{ for unit } i \text{ of } U \text{ and}$$

$$y_{h_i} \ = \ \text{value of } y \text{ for unit } i \text{ of the } h^{th} \text{ stratum}$$

$$\nu_h = \frac{n_h}{n'_h}, (s'_h)^2 = \frac{1}{n'_h - 1} \sum_1^{n'_h} \left(y_{h_i} - \bar{y}'_h\right)^2$$

and
$$S_h^2 = \frac{1}{N_h - 1} \sum_1^{N_h} \left(y_{h_i} - \bar{Y}_h\right)^2, \text{ it follows}$$

$$V(\bar{y}_{dst}) = \left(\frac{1}{n'} - \frac{1}{N}\right) S^2 + \frac{1}{n'} \sum_{h=}^{H} \left(\frac{1}{\gamma_h} - 1\right) W_h S_h^2$$

Chaudhuri (2010) obtained its unbiased estimator

$$v(\bar{y}_{dst}) = \left[\sum_h \left\{ W'_h (\bar{y}_h)^2 - \frac{1}{n'_h} \left(\frac{1}{\gamma_h} - 1\right) s_h^2 \right\} - \bar{y}_{dst}^2 \right] C$$

$$+ \frac{1}{n'} \sum_h \left(\frac{1}{\gamma_h} - 1\right) w'_h s_h^2$$

writing
$$C = \left(\frac{1}{n'} - \frac{1}{N}\right) \frac{N}{N-1} \bigg/ \left\{ 1 - \left(\frac{1}{n} - \frac{1}{N}\right) \frac{N}{N-1} \right\},$$

$$s_h^2 = \left(\frac{1}{n_h} - 1\right) \sum_{i=1}^{n_h} \left(y_{h_i} - \bar{y}_h\right)^2.$$

Chaudhuri (2010) may be consulted for proof.

3.12.2 Double Sampling Ratio Estimation

Suppose an SRSWOR of n' units is taken from $U = (1, \ldots, i, \ldots, N)$ in the first phase and sample means $\bar{y}' = \frac{1}{n'} \sum_1^{n'} y_i$ and $\bar{x}' = \frac{1}{n'} \sum_1^{n'} x_i$ are obtained. Then for $Y = \sum_1^N y_i$, the ratio estimator is

$$\widehat{Y}_R = X \frac{\bar{y}'}{\bar{x}'}.$$

But since $X = \sum_1^N x_i$ may not be known, it cannot be employed in estimation. So, let from the first-phase sample a second-phase SRSWOR of n units be drawn and the sample means $\bar{y} = \frac{1}{n} \sum_1^n y_i$, $\bar{x} = \frac{1}{n} \sum_1^n x_i$ be obtained. Letting $X^* = N\bar{x}'$ a double sampling-based ratio estimator for Y is

$$\widehat{Y}_{dR} = X^* \frac{\bar{y}}{\bar{x}}.$$

Let us further write

$$\widehat{Y} = N\bar{y}, Y^* = N\bar{y}', \widehat{X} = N\bar{x}$$

and

$$e_1 = \frac{\widehat{Y} - Y^*}{Y^*} \Rightarrow \widehat{Y} = Y^*(1 + e_1)$$

$$e_2 = \frac{\widehat{X} - X^*}{X^*} \Rightarrow \widehat{X} = X^*(1 + e_2)$$

and

$$e = \frac{Y^* - Y}{Y} \Rightarrow Y^* = Y(1 + e)$$

So,

$$\widehat{Y}_{dR} = X^* \frac{\bar{y}}{\bar{x}} = X^* \frac{\widehat{Y}}{\widehat{X}}$$

$$= X^* \frac{Y^*(1 + e_1)}{X^*(1 + e_2)} = \frac{Y(1 + e)(1 + e_1)}{(1 + e_2)}$$

Assuming $|e_2| < 1$, we write

$$\widehat{Y}_{dR} \simeq Y(1 + e)(1 + e_1)(1 - e_2 + e_2^2)$$

So,

$$E\widehat{Y}_{dR} \simeq Y E_1 E_2[1]$$

$$+ Y E_1 \left[(1 + e) \left\{ V_2(e_2^2) - \frac{\text{COV}_2(\widehat{X}, \widehat{Y})}{E_2(\widehat{X}) E_2(\widehat{Y})} \right\} \right]$$

So, $\text{Bias}(\widehat{Y}_{dR})$ as an estimator of Y is

$$B(\widehat{Y}_{dR}) \simeq Y \left[\frac{V_2(\widehat{X})}{E_2(\widehat{X})^2} - \frac{\text{COV}_2(\widehat{X}, \widehat{Y})}{E_2(\widehat{X}) E_2(\widehat{Y})} \right].$$

So, an estimate of $B(\widehat{Y}_{dR})$ is

$$b(\widehat{Y}_{dR}) \simeq \widehat{Y} \left[\frac{\widehat{V}_2(\widehat{X})}{(\widehat{X})^2} - \frac{\widehat{\text{COV}}_2(\widehat{X}, \widehat{Y})}{\widehat{X}\widehat{Y}} \right].$$

This, of course, is quite crude and cannot be recommended as good enough. Likewise, the Mean Square Error (MSE) of $\widehat{Y}_{dR}$ as an estimator for Y may be noted as

$$M(\widehat{Y}_{dR}) = Y^2 E_1 E_2 [e\{1 + (e_1 - e_2)\}]^2$$

$$= E_1(Y^* - Y)^2 \left[\frac{V_2(\widehat{Y})}{(Y^*)^2} + \frac{V_2(\widehat{X})}{(X^*)^2} - \frac{2\text{COV}_2(\widehat{Y}, \widehat{X})}{Y^* X^*} \right].$$

So, an estimator for it is

$$m(\widehat{Y}_{dR}) = \widehat{V}_1(Y^*) \left[\frac{\widehat{V}_2(\widehat{Y})}{(Y^*)^2} + \frac{\widehat{V}_2(\widehat{X})}{(X^*)^2} + \frac{2\text{COV}_2(\widehat{Y}, \widehat{X})}{Y^* X^*} \right].$$

This is also not quite serviceable.

3.12.3 Double Sampling for Regression Estimator

Based on a postulated linear regression model writing

$$y_i = \alpha + \beta x_i + \varepsilon_i, i \in U$$

for an SRSWOR of size n from a finite population $U = (1, \ldots, i, \ldots, N)$, a linear regression estimator for $\bar{Y} = \frac{1}{N} \sum_1^N y_i$ is

$$\bar{y}_{lr} = \bar{y} + b(\bar{X} - \bar{x})$$

with
$$\bar{y} = \frac{1}{n} \sum_1^n y_i, \bar{x} = \frac{1}{n} \sum_1^n x_i, \bar{X} = \frac{1}{N} \sum_1^N x_i$$

$$b = \frac{\sum_1^n (y_i - \bar{y})(x_i - \bar{x})}{\sum_1^n (x_i - \bar{x})^2}.$$

The variance of $\bar{y}_{lr}$ is estimated by

$$v(\bar{y}_{lr}) = \left(\frac{1}{n} - \frac{1}{N}\right) s_y^2 (1 - r^2)$$

writing

$$s_y^2 = \frac{1}{n-1} \sum_1^n (y_i - \bar{y})^2 \text{ and}$$

$$r = \frac{\sum_1^n (y_i - \bar{y})(x_i - \bar{x})}{\sqrt{\sum_1^n (y_i - \bar{y})^2} \sqrt{\sum_1^n (x_i - \bar{x})^2}},$$

the sample correlation coefficient between y and x.

If $\bar{X}$ is not available, then $\bar{y}_{lr}$ above is not usable and Double Sampling is needed.

As in the case of ratio estimation in Double Sampling let a first-phase SRSWOR of n' units be drawn from U and $\bar{y}' = \frac{1}{n} \sum_{i \in s'} y_i, \bar{x}' = \frac{1}{n} \sum_{i \in s'} x_i$ be obtained, calling this first-phase sample as s'. Let from this s' an SRSWOR of size $n(< n')$ be drawn calling it the second-phase sample by s and s, s' our Double Sample.

Let

$$b = \frac{\sum_{i \in s'} (y_i - \bar{y})(x_i - \bar{x})}{\sum_{i \in s'} (x_i - \bar{x})^2},$$

$$\bar{y} = \frac{1}{n} \sum_{i \in s} y_i, \bar{x} = \frac{1}{n} \sum_{i \in s} x_i.$$

Now,

$$\bar{y}_{dlr} = \bar{y} + b(\bar{x}' - \bar{x})$$

is taken as the 'Double Sample-based' regression estimator for $\bar{Y}$.

Then, writing

$$B = \frac{\sum_1^N (y_i - \bar{Y})(x_i - \bar{X})}{\sum_1^N (x_i - \bar{X})^2}$$

$$= C_{yx} S_x^2,$$

$$S_x^2 = \frac{1}{N-1} \sum_1^N (x_i - \bar{X})^2 \text{ and}$$

$$C_{yx} = \frac{1}{N-1} \sum_1^N (y_i - \bar{Y})(x_i - \bar{X}).$$

Then approximately,

$$\bar{y}_{dlr} \simeq \bar{y} + B(\bar{x}' - \bar{x}) = \bar{y}'_{dlr}, \text{ say.}$$

Writing
$$u_i = y_i - B(x_i), \bar{u} = \frac{1}{n'} \sum_{i \in s'} u_i$$

$$s'^2_u = \frac{1}{(n'-1)} \sum_{i \in s'} (u_i - \bar{u})^2.$$

Then,
$$V_2(\bar{y}'_{dlr}) \simeq \left(\frac{1}{n} - \frac{1}{n'}\right)(s'_u)^2.$$

With simple algebra following earlier works,

$$V(\bar{y}'_{dlr}) \simeq \left(\frac{1}{n'} - \frac{1}{N}\right) s_y^2 + \left(\frac{1}{n} - \frac{1}{n'}\right) \frac{1}{(N-1)} \sum_1^N [(y_i - \bar{Y}) - B(x_i - \bar{X})]^2.$$

A simple estimator for it is derived and given by Chaudhuri (2010) as

$$v(\bar{y}'_{dlr}) = \frac{N^2(1-f)}{n} \frac{1}{(n-1)} \sum_{i \in s} (e_i - \bar{e})^2$$

writing
$$f = \frac{n}{N}, e_i = y_i - bx_i \text{ and}$$

$$\bar{e} = \frac{1}{n} \sum_{i \in s} e_i, b = \frac{\sum_{i \in s} (y_i - \bar{y})(x_i - \bar{x})}{\sum_{i \in s} (x_i - \bar{x})^2}$$

3.12.4 Double Sampling in Tackling Non-response

Suppose from a finite population $U = (1, \ldots, i, \ldots, N)$ of N units an SRSWOR of $n(< N)$ units is taken. But in the survey, only $n_1(< n)$ of them give data on a variable y of interest. This is called a 'Non-Response' problem. Then out of the $n_2 = n - n_1$ non-respondents let an SRSWOR of $r(< n_2)$ units be freshly taken and suppose luckily each of them answers. This is also a case of Double sampling.

Suppose $\bar{y}_1 = \frac{1}{n_1} \sum_1^{n_1} y_i, \bar{y}_{2r} = \frac{1}{r} \sum_1^r y_i$ are the means of the y_i-values actually observed and $\bar{y}_2 = \frac{1}{n_2} \sum_1^{n_2} y_i$ be the mean we could not calculate for the y_i-values of the n_2 missing units that we needed to obtain but could not.

Let us now consider
$$\bar{y}' = \frac{n_1 \bar{y}_1 + n_2 \bar{y}_{2r}}{n_1 + n_2}$$

as an estimator for $\bar{Y} = \frac{1}{N} \sum_1^N y_i.$

As usual denoting by E_1, E_2, E and V_1, V_2, V the operators for expectation and variance for the initial, subsequent and the over-all double sampling as above, we may note following:

$$E_2(\bar{y}_{2r}) = \bar{y}_2$$

$$E_2(\bar{y}') = \frac{n_1}{n}\bar{y}_1 + \frac{n_2}{n}\bar{y}_2 = \bar{y} = \frac{1}{n}\sum_1^n y_i$$

though $\bar{y}$ is not observed. Now,

$$V(\bar{y}') = V_1 E_2(\bar{y}') + E_1 V_2(\bar{y}')$$
$$= V_1(\bar{y}) + E_1\left[\left(\frac{n_2}{n}\right)^2\left(\frac{1}{r} - \frac{1}{n_2}\right)\right]S_2^2$$
$$= \left(\frac{1}{n} - \frac{1}{N}\right)S^2 + W_2(K-1)\frac{S_2^2}{n}$$

writing

$$W_1 = \frac{N_1}{N}, W_2 = \frac{N_2}{N}, K = \frac{n_2}{r}$$

$$S_2^2 = \frac{1}{(N_2-1)}\sum_1^{N_2}\left(y_i - \frac{\sum_1^{N_2}y_i}{N_2}\right)^2, S^2 = \frac{1}{N-1}\sum_1^N\left(y_i - \frac{\sum_1^N y_i}{N}\right)^2$$

writing N_1 and N_2 as the assumed numbers of units in the population to be available and unavailable to answer in the first attempted survey.

This $V(\bar{y}')$, though cannot be determined, provides an interesting clue to answer reasonably the two questions, namely how many units to be sampled first and later on encountering a few non-respondents, of course, on choosing a cost function

$$C' = C_0 n + C_1 n_1 + C_2 r$$

take C_0, C_1 and C_2 as costs per unit of choosing the initial SRSWOR of size n, surveying the n_1 respondents first addressed and those surveyed the 2nd draw of samples.

Let us write

$$V = V(\bar{y}') + \frac{S^2}{N} = \frac{1}{n}(S^2 - W_2 S_2^2) + \frac{K W_2 S_2^2}{n}$$

and

$$C = E(C') = n(C_0 + C_1 W_1) + \frac{C_2}{K}n W_2.$$

Then,

$$VC = \alpha + \beta K + \frac{\theta}{K}, \text{ on writing}$$

$$\alpha = (S^2 - W_2 S_2^2)(C_0 + C_1 W_1) + C_2 W_2^2 S_2^2$$
$$\beta = (W_2 S_2^2)(C_0 + C_1 W_1), \quad \theta = C_2 W_2 (S^2 - W_2 S_2^2)$$

A way to optimally choose n and K is either to fix 'a' C and minimize V subject to this specification or to fix 'a' V and minimize C subject to this fixation. For this, the first step is to minimize VC with respect to 'K'.

Now
$$VC = \alpha + \beta K + \frac{\theta}{K};$$

$$\beta K + \frac{\theta}{K} = (\sqrt{\beta K})^2 + \left(\sqrt{\frac{\theta}{K}}\right)^2$$
$$= \left(\sqrt{\beta K} - \sqrt{\frac{\theta}{K}}\right)^2 + 2\sqrt{\beta\theta}.$$

This is minimum for the choice of K as $K_{\mathrm{opt}} = \sqrt{\frac{\theta}{\beta}}$.
So, the minimum value is

$$VC = \alpha + \beta K_{\mathrm{opt}} + \frac{\theta}{K_{\mathrm{opt}}}.$$

For a fixed V or a fixed C, the optimum value of n then

is
$$n_{\mathrm{opt}} = \left[(S^2 - W_2 S_2^2) + W_2 S_2^2 K_{\mathrm{opt}}\right] \Big/ C$$

or is
$$n_{\mathrm{opt}} = \frac{C}{(C_0 + C_1 W_1) + \frac{C_2 W_2}{K_{\mathrm{opt}}}}.$$

Hansen and Hurwitz (1946) introduced this approach which is, of course, an intellectual exercise because the solution cannot be implemented because so many unknowable parameters are involved.

3.13 Sampling on Successive Occasions

Suppose a population $U = (1, \ldots, i, \ldots, N)$ with values $\underline{X} = (x_1, \ldots, x_i, \ldots, x_N)$, $X = \sum_1^N x_i$ on a nearby preceding occasion for a variable x on the current occasion has values $\underline{Y} = (y_1, \ldots, y_i, \ldots, y_N)$, $Y = \sum_1^N y_i$ for the same variable but we denote it differently as y because of the slight advance in time. Our problem is to suitably estimate the current population total Y on supposing that on the immediate preceding occasion a sample s_1 was chosen and surveyed in respect of $\underline{X}$; next on the neighbouring current occasion a part of s_1 called a 'matching sample' s_m is taken out of s_1 and surveyed in respect of $\underline{Y}$ and a further sample called 'unmatched sample' s_u is drawn either from U or from $U - s_1$ or from $U - s_m$ and is surveyed on $\underline{Y}$. Now a suitable estimate for Y is to be employed utilizing the values of x_i for $i \in s_1$ and y_i for i in s_m and s_u. Such an estimate should be of the form

$$t = \phi[t_1(s_1, x_i | i \in s_1) + t_2(s_m, x_i, y_i | i \in s_m)]$$
$$+ (1 - \phi)t_3(s_u, y_i | i \in s_u)$$

with suitable choice of forms for t_1, t_2, t_3 and a combining weight ϕ with appropriate choice of the matching fraction $\frac{m}{n_1}$.

The topic 'Sampling on Successive Occasions' has a long and broad history. Jessen (1942) stated it with his estimator

$$t = \phi[\bar{x}_1 + b(\bar{y}_m - \bar{x}_m)] + (1 - \phi)\bar{y}_u$$

taking an initial SRSWOR s_1 of size n, a matched SRSWOR of size $m(< n)$ from s_1 and an unmatched sample s_u of size n from U. The means of y, x from s_1, s_m and s_u are used in t with obvious notations and

$$b = \frac{\sum_{s_m}(y_i - \bar{y}_m)(x_i - \bar{x}_m)}{\sum_{s_m}(x_i - \bar{x}_m)^2}.$$

The stress was on working out m and ϕ in appropriate ways.

Yates (1949), Patterson (1950), Tikkiwal (1953) and Eckler (1955) among others extended this to sampling on more than two occasions. Kulldorff (1963) used s_u chosen from $U - s_1$ and (Singh, 1972) chose s_u from $U - s_m$. Pathak and Rao (1967) modified t by Rao-Blackwellization noting common units between s_m and s_u, if the latter is chosen not from $U - s_m$. Kulldorff (1963) and Sen (1973) used different sizes of s_1 and s_u.

Raj (1965) chose s_1 and s_u by PPSWR using some known size-measures but s_m by SRSWOR method. Obviously, his t_1 and t_3 were Hansen and Hurwitz's estimators.

Ghangurde and Rao (1969) chose s_1 and s_u from U by Rao et al. (1962) method and s_m by SRSWOR. Chotai (1974) took s_m also by RHC method. Avadhani and Sukhatme (1970) chose s_m from s_1 by RHC scheme utilizing x_i for $i \in s_1$. Further alternatives were also employed by many others.

Chaudhuri and Arnab (1977, 1979, 1982) modified earlier procedures by introducing the Lahiri, Midzuno and Sen sampling scheme and the Horvitz-Thompson estimator and suggested estimating Y by

$$e = \frac{a}{m} \sum_{s_m} \frac{y_i}{z_i} + \frac{b}{m} \sum_{s_n} \frac{x_i}{z_i}$$

$$+ \frac{c}{n} \sum_{1}^{n} \frac{x_r}{z_r} + \frac{d}{n} \sum_{r=1}^{n} \frac{y_r}{z_r}$$

with a, b, c, d to choose suitably. Chaudhuri and Adhikary (1983) discussed on the efficiency of Midzuno's and Sen's strategy relative to some ratio type estimators.

Rao and Bellhouse (1978) and Chaudhuri (1985) gave several sophisticated further developments in this context of sampling on two nearby timepoints presuming a fixed population for both.

Chaudhuri (2014) may be profitably consulted for a rather comprehensive narration.

3.14 Panel Rotation Sampling

In the context of sampling on several consecutive occasions, two terms are crucial. One is 'Panel'. A panel means a set of units to be sampled from a population which on survey are to report about a variable of interest for a specific point or period of time. The other is 'Rotation'. By 'rotation' is meant 'retaining' a portion of a sample taken and surveyed on one occasion and 'dropping' or 'omitting' the remaining units from the panel and 'replacing' them by a complement of as many units on the next occasion.

Rao and Graham (1964) are among the pioneers in this research activity among others, like (Patterson, 1950; Eckler, 1955) in this type of ventures, including (Hansen et al., 1955) among others.

Binder and Hidiroglou (1988) gave us a wealth of information in respect of sample surveys across time. Chaudhuri (2010, 2013, 2014, 2018) provides some specifics, we intend to cover here in brief.

In Canadian Monthly Labour Force surveys, one-sixth of the sample households is replaced each by a new set of households each, and each sample is composed of six panels.Each panel continues to report for 6 consecutive months each and then are rotated out the monthly Consumer Population Surveys (CPS) in US the system in operation is 4–8–4. A panel reports for the first 4 months, then drops out for the next 8 months and returns for the next 4 months. These panels constitute a cycle and the survey continues in cycles.

Panel Surveys are a specific aspect of what are known as Longitudinal surveys. In longitudinal surveys in respect of the same sampling units, data are gathered on successive points of time to take account of how they change over time. By contrast in Cross-sectional surveys, data for various sampling units are gathered on a fixed point of time. In medical studies, longitudinal surveys are especially applicable with profit. In this text, we may remain satisfied with an understanding of Panel Surveys instead of going deeper into wider aspects of longitudinal surveys in a further generality

Chaudhuri (2010, pp. 137–140) has comprehensively described the concepts of 'Permanent Random Numbers' which are very useful to draw SRSWORs from finite populations which from time to time change in compositions and yet it is possible to set numbers of units in samples from draw to draw to remain common with specified proportions. Chaudhuri (2010, pp. 137–140) should be consulted to master this particular topic. Here besides SRSWOR, the only varying probability sampling one may resort to is only Poisson Sampling and a few modifications especially collocated sampling one should pay attention to.

3.15 Lessons

(i) Given $U = (1, 2, \ldots, 10)$ with $\underline{X} = (x_1, x_2, \ldots, x_{10})$ and a PPSWR sample $s = (6, 9, 4, 4, 1, 9)$ with y-values $\underline{y}_s = (3.8, 1.7, 3.3, 3.3, 5.0, 1.7)$, find an unbiased estimate for $Y = \sum_1^{10} y_i$, $\underline{Y} = (y_1, y_2, \ldots, y_{10})$ and an unbiased estimator for the variance of the values of y_i, $i = 1, \ldots, 10$.

Solution:

An unbiased estimator for $Y = \sum_1^{10} y_i$

$$t_{HH} = \frac{1}{6}\left[\frac{3.8}{x_6} + 2\frac{1.7}{x_9} + 2\frac{3.3}{x_4} + \frac{5.0}{x_1}\right]X,$$

with $X = \sum_1^{10} x_i$ and $p_i = \frac{x_i}{X}, i \in U.$

$$\widehat{V}(t_{HH}) = \frac{1}{2 \times 6 \times 5}\left[\sum_{r \neq r'=1}^{6}\sum^{6}\left(\frac{y_r}{p_r} - \frac{y_{r'}}{p_{r'}}\right)^2\right]$$

$$E\widehat{V}(t_{HH}) = \frac{1}{6}\left[\sum_1^N \frac{y_i^2}{p_i} - Y^2\right]$$

$$\widehat{Y}^2 = \frac{1}{6}\left[\sum_{r=1}^6 \frac{y_r^2}{p_r}\frac{1}{6p_r}\right] - \widehat{V}(t_{HH})$$

$$V(\bar{y}) = \frac{1}{9}\left[\sum_1^{10} y_i^2 - \frac{Y^2}{10}\right]$$

$$\widehat{V}(\bar{y}) = \frac{1}{9}\left[\frac{1}{6}\sum_{r=1}^6 \frac{y_r^2}{p_r} - \frac{\widehat{Y}^2}{10}\right].$$

(ii) Suppose $y_{ijk} \equiv$ the value of a real variable y for the kth ($k = 1, \ldots, B_{ij}$) third-stage unit in the jth ($j = 1, \ldots, M_i$) second-stage unit of the ith ($i = 1, \ldots, N$) first-stage unit of a finite population.

Let $p_i(0 < p_i < 1, \sum_1^N p_i = 1)$ be known normed size-measures for the first-stage units.

(a) Take a PPSWR sample of first-stage units in $n(2 \le n < N)$ draws; from each fsu whenever drawn take an SRSWOR of 2 ssus out of all the ssus in it and from each ssu so drawn take an SRSWOR of 3 tsus among those in the corresponding ssu sampled. Obtain an unbiased estimator of the population total

$$Y = \sum_1^N \sum_{j=1}^{M_i} \sum_{k=1}^{B_{ij}} y_{ijk}$$

along with an unbiased estimator for the variance of this three-stage estimator. Repeat this exercise taking the sample of the fsus by PPSWOR method and employing Des Raj's estimation method in the first stage keeping intact everything else.

Solution:

Let a PPSWR sample of fsus be taken in n draws as

$$s = (i_1, \ldots, i_r, \ldots, i_n)$$

with y_r as the y-value for the unit chosen on the rth draw ($r = 1, \ldots, n$). But y_r cannot be determined but is supposed to be estimated unbiasedly by

$$\widehat{y_r} = \sum_{j=1}^{m_r} \frac{M_r}{m_r} \left(\frac{B_{rj}}{b_{rj}} \sum_{k=1}^{b_{rj}} y_{rjk} \right)$$

writing $B_{rj}(b_{rj})$ as the numbers of tsus in the ssus (sampled out of them by SRSWOR) in the fsu chosen on the rth draw; further, M_r is the number of ssus in the fsu chosen on the rth draw and m_r is the number of ssus chosen out of them by SRSWOR. Here, b_{rj} is 3 and m_r is 2 as are specified here.

An unbiased estimator for Y is

$$\widehat{Y} = \frac{1}{n} \sum_{r=1}^n \left(\widehat{y_r}/p_r \right)$$

because $E(\widehat{Y}) = E_p \left[E_L \left(\frac{1}{n} \sum_{r=1}^n \widehat{y_r}/p_r \right) \right]$

$$= E_p \left(\frac{1}{n} \sum_{r=1}^{n} \frac{y_r}{p_r} \right) = \sum_{1}^{N} y_i = Y.$$

Also,

$$V(\widehat{Y}) = E_p V_L(\widehat{Y}) + V_p E_L(\widehat{Y})$$

$$= E_p V_L(\widehat{Y}) + V_p \left(\frac{1}{n} \sum_{r=1}^{n} \frac{y_r}{p_r} \right)$$

$$= E_p V_L(\widehat{Y}) + \frac{1}{2n} \sum_{1}^{N} \sum_{1}^{N} \left(\frac{y_i}{p_i} - \frac{y_j}{p_j} \right)^2 p_i p_j$$
$$\underset{i \neq j}{}$$

Let

$$\nu(\widehat{Y}) = \frac{1}{2n^2(n-1)} \sum_{\substack{r,r'=1 \\ r \neq r'}}^{n} \sum_{}^{n} \left(\frac{\widehat{y_r}}{p_r} - \frac{\widehat{y_{r'}}}{p_{r'}} \right)^2$$

Then,

$$E_L \nu(\widehat{Y}) = \frac{1}{2n^2(n-1)} \sum_{\substack{1 \\ r \neq r'}}^{n} \sum_{1}^{n} E_L \left(\frac{\widehat{y_r}}{p_r} - \frac{\widehat{y_{r'}}}{p_{r'}} \right)^2$$

$$= \frac{1}{2n^2(n-1)} \sum_{\substack{r=1 \\ r \neq r'}}^{n} \sum_{r'=1}^{n} \left[\sum_{r=1}^{n} \sum_{r'=1}^{n} \left(\frac{y_r}{p_r} - \frac{y_{r'}}{p_{r'}} \right)^2 \right.$$

$$\left. + \sum_{r=1}^{n} \sum_{r'=1}^{n} \left[V_L \left(\frac{\widehat{y_r}}{p_r} \right) + V_L \left(\frac{\widehat{y_{r'}}}{p_{r'}} \right) \right] \right]$$

Then,

$$E\nu(\widehat{Y}) = E_p \left[E_L V(\widehat{Y}) \right]$$

$$= E_p V_L(\widehat{Y}) + \frac{1}{2n} \sum_{i=1}^{N} \sum_{j=1}^{N} \left(\frac{y_i}{p_i} - \frac{y_j}{p_j} \right)^2 p_i p_j$$
$$\underset{i \neq j}{}$$

$$= V(\widehat{Y})$$

Thus, $\nu(\widehat{Y})$ above is an unbiased estimator for $V(\widehat{Y})$.
In the second case where the n fsus are chosen by PPSWOR method, the following changes are needed.
If an fsu i is selected, its y-value y_i cannot be determined but it is expressible as

$$y_i = \sum_{j=1}^{M_i} \sum_{k=1}^{B_{ij}} y_{ijk}, i \in s;$$

s is a PPSWOR sample of fsus.
This y_i is then unbiasedly estimated by

$$\widehat{y_i} = \frac{M_i}{m_i} \sum_{j=1}^{m_i} \frac{B_{ij}}{b_{ij}} \sum_{k=1}^{b_{ij}} y_{ijk}$$

$$= \frac{M_i}{m_i} \sum_1^{m_i} \widehat{Y_{ij}}, \text{ say.}$$

Then, $\qquad E_3\left(\frac{B_{ij}}{b_{ij}} \sum_{k=1}^{b_{ij}} y_{ijk}\right) = \sum_{k=1}^{B_{ij}} y_{ijk} = Y_{ij}, \text{ say.}$

$$E_2\left(\frac{M_i}{m_i} \sum_{j=1}^{m_i} Y_{ij}\right) = \sum_{j=1}^{M_i} y_{ij} = Y_i, \text{ say.}$$

Finally, $\qquad E(\widehat{y_i}) = E_1(Y_i).$

But $\widehat{Y} = \widehat{Y}_{DR}$ = Des Raj's unbiased estimator of $Y = \sum_1^N y_i$ to be obtained from Y_i using the n fsus sampled from $U = (1, \ldots, i, \ldots, N)$ by PPSWOR sampling method.
Unbiased estimator of $V(\widehat{Y}_{DR})$ is to be obtained as follows:

$$\widehat{V}(\widehat{y_i}) = \frac{M_i - m_i}{M_i m_i} \frac{1}{(m_i - 1)} \sum_1^{m_i} \left(\widehat{Y}_{ij} - \frac{\sum_1^{b_{ij}} \widehat{Y}_{ij}}{b_{ij}}\right)^2$$

$$+ \frac{B_{ij} - b_{ij}}{B_{ij} b_{ij}} \frac{1}{(b_{ij} - 1)} \sum_1^{b_{ij}} \left(y_{ijk} - \frac{\sum_1^{b_{ij}} y_{ijk}}{b_{ij}}\right)^2.$$

Finally,

$$V(\widehat{Y}) = \frac{1}{2n^2(n-1)} \sum_{i \neq i'}^n \sum^n (\widehat{y_i} - \widehat{y_{i'}})^2 + \frac{N}{n} \sum_{i=1}^n \widehat{V}(\widehat{y_i}).$$

(iii) In stratified sampling with SRSWOR independently across the strata, find a rule to determine the over-all sample-size.
Solution:
On p11 a rule is provided to specify the sample-size n to draw from a finite population of size N by SRSWOR as

$$n = N/\{1 + \alpha N f^2 (100)^2/(\text{CV})^2\}, \, vide(1.2.3).$$

Here, $\text{Prob}[|\bar{y} - \bar{Y}| \leq f\bar{Y}] \geq 1 - \alpha$ with f and α as two positive proper fractions and $CV = 100S/\bar{Y}$ (S is the population standard deviation of the values of y for the N population units).

Now apply this formula (1.4.3) to obtain a rule to work out the sample-size to independently draw from each of the H strata by SRSWOR. The sum of these stratum-wise sample-sizes gives the over-all n of the sample from the population.

Since (1.4.3) is obtained by using Chebyshev's inequality, we may regard the stratum-wise sample-sizes derived above as the allocation of the over-all sample-size n to the strata by applying Chebyshev's inequality.

(iv) By arbitrarily choosing H and the strata-wise CVs and the population means and hence deriving the S-values, evaluate the variances of the standard unbiased estimator for the population mean by stratified sampling on employing the allocation rules, namely equal, proportional, Neyman's optimal allocation rules and also the one by Cheshev's rule and comment. In an unpublished paper, Chaudhuri and Chandrima Chakraborty have done this exercise and found Neyman's rule to give the best result followed by Chebyshev's rule, proportional and equal allocation rule in this order as it should be the case.

References

Ajgaonkar, S. G. P. (1967). Unbiased estimates of the variance of the Narain, Horvitz and Thompson estimator. *Sankhyā B, 29*, 55–60.

Avadhani, M. S., & Sukhatme, B. V. (1970). A comparison of two sampling procedures with an application to successive sampling. *JRSS C, 19*, 231–259.

Binder, D. A., & Hidiroglou, M. A. (1988). Sampling in time. In P. R. Krishnaiah & C. R. Rao (Eds.), *Handbook of statistics* (Vol. 6, pp. 187–211). NH, Amsterdam.

Chaudhuri, A., & Arnab, R. (1977). On the relative efficiencies of a few strategies of sampling with varying probabilities on two occasions. *CSA Bulletin, 26*, 25–38.

Chaudhuri, A., & Arnab, R. (1979). On the relative efficiencies of sampling strategies under a super-population model. *Sankhyā C, 41*(1), 40–43.

Chaudhuri, A., & Arnab, R. (1982). On unbiased variance estimators with various multi-stage sampling strategies. *Sankhyā B, 44*, 92–101.

Chaudhuri, A., & Adhikary, A. K. (1983). On the efficiency of Midzuno and Sen's strategy relative to several ratio-type estimators under a particular model. *Biometrika, 79*(2), 121–124.

Chaudhuri, A. (1985). An optimal and related strategies for sampling on two occasions with varying probabilities. *JISAS, 37*, 45–53.

Chaudhuri, A. (2010). *Essentials of survey sampling*. New Delhi, India: PHI.

Chaudhuri, A. (2013). Panel rotation with general sampling schemes. *JISAS, 67*(3), 301–304.

Chaudhuri, A. (2014). *Modern survey sampling*. Florida, USA: CRC Press.

Chaudhuri, A. (2018). *Survey Sampling*. Florida, USA: CRC Press.

Chaudhuri, A., & Pal, S. (2003). On a version of cluster sampling and its practical use. *JSPI, 113*, 25–34.

Chaudhuri, A., & Pal, S. (2003). Systematic sampling: "Fixed" versus "Random" sampling interval. *PJS, 19*(2), 259–271.

Chotai, J. (1974). A note on Rao-Hartley-Cochran method for PPS over two occasions. *Sankhya, Ser C.*, 36, 173–180.

Dalenius, T. (1950). The problem of optimum stratification. *Scandinavian Actuarial Journal*, 203–213.

Das, M. N. (1982). Systematic sampling without drawback, Technical report, 8206, ISI, Delhi

Deming, W. E. (1956). On simplification of sampling design through replication with equal probabilities and without stages. *JASA, 51*, 24–53.

Raj, Des. (1956). Some estimators in sampling with varying probabilities without replacement. *JASA, 274*, 269–284.

Eckler, A. R. (1955). Rotation sampling. *AMS, 26*, 664–685.

Doss, D. C., Hartley, H. O., & Somayajulu, G. R. (1979). An exact small sample theory for post-stratification. *JSPI, 3*, 235–248.

Ghangurde, P. D. & Rao, J. N. K. (1969). Some results on sampling over two occasions. *Sankhya, Ser A.*, 31, 463–472.

Hansen, M. H., & Hurwitz, W. N. (1943). On the theory of sampling from finite populations. *The Annals of Mathematical Statistics, 14*, 333–362.

Hansen, M. H., & Hurwitz, W. N. (1946). The problem of non-Response in sample surveys. *JASA, 41*, 517–529.

Hansen, M. H., Hurwitz, W. N., Nisselson, H., & Steinberg, J. (1955). The redesign of the census current population survey. *JASA, 50*, 701–719.

Horvitz, D. G., & Thompson, D. J. (1952). A generalization of sampling without replacement from a finite universe. *JASA, 47*, 663–689.

Jessen, R. J. (1942). *Statistical investigation of a sample durey for obtaining farm facts* (p. 304). Bull: Iowa Agri. Exp. Stn. Res.

Koop, J. C. (1960). On theoretical questions underlying the technique of replicated or interpenetrating samples. In *Proceedings of the Social Statistics Section* (pp. 196–205). American Statistical Association.

Kulldorff, G. (1963). Some problems of optimal allocation for sampling on two occasions. *ISR, 31*, 24–57.

Lahiri, D. B. (1951). A method of sample selection providing unbiased ratio estimators. *BISI, 33*(2), 133–146.

Lahiri, D. B. (1954). Technical paper on some aspects of the development of the sample design: National Sample Survey Report no. 5. *Sankhyā, 14*, 264–316.

Mahalanobis, P. C. (1939). First report of the crop census of 1938, Indian Central Jute Committee

Midzuno, H. (1950). An outline of the theory of sampling systems. *Annals of the Institute of Statistical Mathematics, 1*, 149–156.

Murthy, M. N. (1957). Ordered and unordered estimators in sampling without replacement. *Sankhyā, 18*, 379–390.

Narain, R. D. (1951). On sampling without replacement with varying probabilities. *JISAS, 3*, 169–175.

Neyman, J. (1934). On the two different aspects of the representative method, the method of stratified sampling and the method of purposive selection. *JRSS, 97*, 558–625.

Neyman, J. (1938). Contributions to the theory of sampling human populations. *JASA, 33*, 101–116.

Patterson, H. D. (1950). Sampling on successive occasions with partial replacement of units. *JRSS B, 12*, 241–255.

Pathak, P. K., & Rao, T. J. (1967). Inadmissibility of customary estimators in sampling over two occasions. *Sankhyā, 29*, 49–54.

Raj, Des. (1965). On sampling over two occasions with probability proportionate to size. *AMS, 36*, 327–330.

Rao, J. N. K., & Graham, J. R. (1964). Rotation designs for sampling on repeated occasions. *JASA, 59*, 492–509.

Rao, J. N. K., Hartley, H. O., & Cochran, W. G. (1962). On a simple procedure of unequal probability sampling without replacement. *JRSS B, 24*, 482–491.

Rao, J. N. K., & Bellhouse, D. R. (1978). Optimal estimation of a finite population mean under generalized random permutation models. *JSPI, 2,* 125–141.

Roychaudhury, D. K. (1957). Unbiased sampling design using information provided by linear function of auxiliary variate, Chapter 5, thesis for Associateship of Indian Statistical Institute, Kolkata.

Sen, A. R. (1973). Some theory of sampling on successive occasions. *AJS,* 15, 105–110.

Singh, R. (1972). On Pathak and Rao's estimate in pps with replacement sampling over two occasions. *Sankhyā, A, 34,* 301–303.

Sudhakar, K. (1978). A note on circular systematic sampling design. *Sankhya C, 40,* 72–73.

Tikkiwal, B. D. (1953). Optimum allocation in successive sampling. *JISAS, 5*(1), 100–102.

Yates, F. (1949). *Sampling methods for census and surveys.* London: Griffin & Co.

Yates, F., & Grundy, P. M. (1953). Selection without replacement from within strata with probability proportional to size. *Journal of the Royal Statistical Society, 15,* 253–261.

Chapter 4
Fixing the Size of an Equal Probability Sample

4.1 Sample Size Determination in SRSWR

Suppose from a population $U = (1, \ldots, i, \ldots, N)$ of size N an SRSWR of size n is to be chosen to unbiasedly estimate the population mean $\bar{Y} = \frac{1}{N} \sum_{1}^{N} y_i$ of a variable y with values y_i for $i \in U$ employing the sample mean $\bar{y}$ as an estimator. Suppose for its accuracy we like that the error $e = (\bar{y} - \bar{Y})$ be required to be so controlled that $|\bar{y} - \bar{Y}| \leq f\bar{Y}$ with the fraction $f (0 < f < 1)$ so stipulated that, for example, f should be 0.10, i.e., 10% and this contingent may be achieved such that

$$\text{Prob}\left(|\bar{y} - \bar{Y}| \leq f\bar{Y}\right) \geq 1 - \alpha \qquad (4.1.1)$$

with $\alpha(0 < \alpha < 1)$ as a positive proper fraction, say, as low as $\alpha = 0.05$ or 5%. Thus, we demand the sample size to be at least so large that the sample mean $\bar{y}$ in accuracy should be so attained that the error on either side of the population mean be within a fraction f of the true unknown mean $\bar{Y}$ and this should be achieved with at least a $100(1 - \alpha)$ percent chance.

Chebyshev's inequality gives us

$$\text{Prob}\left(|\bar{y} - \bar{Y}| \leq \lambda\sigma(\bar{y})\right) \geq 1 - \frac{1}{\lambda^2} \qquad (4.1.2)$$

writing $\sigma(\bar{y}) = \sqrt{V(\bar{y})}$, the standard deviation of $\bar{y}$ and λ is a positive constant.

Comparing (4.1.1) with (4.1.2) and letting $\alpha = \frac{1}{\lambda^2}$ and $f\bar{Y} = \lambda\sigma(\bar{y})$, we may derive

© The Author(s), under exclusive license to Springer Nature Singapore Pte Ltd. 2022
A. Chaudhuri and S. Pal, *A Comprehensive Textbook on Sample Surveys*, Indian Statistical Institute Series, https://doi.org/10.1007/978-981-19-1418-8_4

$$\alpha = \frac{1}{f^2} \frac{V(\bar{y})}{(\bar{Y}^2)} = \frac{1}{f^2} \left(\frac{CV(\bar{y})}{100}\right)^2$$

$$\text{i.e., } 100f = \frac{CV(\bar{y})}{\sqrt{\alpha}} \tag{4.1.3}$$

recalling that the coefficient of variation of $\bar{y}$, is $CV(\bar{y}) = 100 \frac{\sigma(\bar{y})}{\bar{Y}}$.

For SRSWR in n draws

$$V(\bar{y}) = \frac{1}{Nn} \sum_{i=1}^{N} (y_i - \bar{Y})^2 = \frac{\sigma^2}{n}, \text{ with}$$

$$\sigma^2 = \frac{1}{N} \sum_{1}^{N} (y_i - \bar{Y})^2. \text{ Hence we demand}$$

$$\alpha = \frac{1}{(100f)^2} (100)^2 \frac{\sigma^2}{n(\bar{Y})^2}$$

$$= \frac{1}{nf^2} \left(\frac{N-1}{N}\right) \frac{S^2}{(\bar{Y}^2)}, \text{ writing}$$

$$S^2 = \frac{1}{N-1} \sum_{1}^{N} (y_i - \bar{Y})^2 = \frac{N}{N-1} \sigma^2 \text{ or } \sigma^2 = \frac{N-1}{N} S^2.$$

We shall write for convenience $CV = 100 \frac{S}{\bar{Y}}$, the coefficient of variation of all the N values of y_i in the population, $i \in U$.

So, for SRSWR, to estimate $\bar{Y}$ by $\bar{y}$ the rule is

$$n = \frac{N-1}{N} \frac{(CV)^2}{(100\alpha f^2)} \tag{4.1.4}$$

rounded up to the nearest integer.

Since for SRSWOR in n draws the sample mean $\bar{y}$ unbiasedly estimates $\bar{Y}$ with a variance

$$V(\bar{y}) = \frac{N-n}{Nn} S^2 = \left(\frac{1}{n} - \frac{1}{N}\right) S^2$$

and so, the rule is

$$n = \frac{N}{1 + N\alpha \left(\frac{100f}{CV}\right)^2} \tag{4.1.5}$$

rounded up to the nearest integer because

$$CV(\bar{y}) = 100\left(\frac{S}{\bar{Y}}\right)\sqrt{\left(\frac{1}{n} - \frac{1}{N}\right)} = CV\sqrt{\left(\frac{1}{n} - \frac{1}{N}\right)} \text{ and}$$

$$\alpha = \frac{1}{(100f)^2}(CV)^2\left(\frac{1}{n} - \frac{1}{N}\right) \text{ using (4.1.3)}.$$

More generally, if to estimate a population parameter θ an unbiased estimator e is to be based on a sample chosen according to a design p we may need instead of (4.1.1), the condition

$$Prob(|e - \theta| \le f\theta) \ge 1 - \alpha \qquad (4.1.6)$$

and take recourse to the Chebyshev inequality

$$Prob(|e - \theta| \le \lambda\sigma_p(t)) \ge 1 - \frac{1}{\lambda^2}, \qquad (4.1.7)$$

writing $V_p(t)$ as the variance of t and $\sigma_p(t) = +\sqrt{V_p(t)}$, its standard deviation.

In this however, we should confine only to SRSWR and SRSWOR and postpone to a later chapter (Chap. 13) the treatment of alternative designs while setting the sample sizes.

Let us take θ as $\bar{Y}$, but e as the mean $\bar{y}_d$ based on the d distinct units in an SRSWR in n draws from U.

Then, we know, vide (Chaudhuri, 2010 pp. 35–36), that

$$V(\bar{y}_d) = \frac{1}{N}\left[\sum_{j=1}^{N}\left(\frac{j}{N}\right)^{n-1} - 1\right]S^2.$$

So, combining (4.1.6) and (4.1.7) above we need

$$\alpha = \frac{1}{\lambda^2},$$

$$f\bar{Y} = \lambda\left[\frac{1}{N}\left\{\sum_{1}^{N}\left(\frac{j}{N}\right)^{n-1} - 1\right\}\right]^{\frac{1}{2}} S \text{ and hence}$$

$$100f\sqrt{\alpha} = \left[\frac{1}{N}\left\{\sum_{1}^{N}\left(\frac{j}{N}\right)^{n-1} - 1\right\}\right]^{\frac{1}{2}} CV \qquad (4.1.8)$$

with $CV = 100\frac{S}{\bar{Y}}$.

So, the required sample size n should be tabulated in terms of N, f, α and CV and referring to the formula (4.1.8), to try to solve this we start with n from (4.1.4) and putting that n in (4.1.8), try to adjust keeping the LHS and RHS as close as possible and rounded up to the nearest integer.

Next we consider θ as Y but a few alternative unbiased estimators for Y based on SRSWOR as e, the Hartley and Ross (1954) estimator

$$e_{HR} = N\left[\bar{r} + \left(\frac{N-1}{N}\right)\left(\frac{n}{n-1}\right)\frac{(\bar{y} - \bar{r}\bar{x})}{\bar{X}}\right] \tag{4.1.9}$$

with $\bar{r} = \frac{1}{n}\sum_{i\in s}\frac{y_i}{x_i}$, $\bar{x}$, $\bar{y}$ the sample means of y and another correlated variable x with

respective values x_i for i in a sample s by SRSWOR from U and $\bar{X} = \frac{1}{N}\sum_{i=1}^{N}x_i$.

The $V(e_{HR})$ is complicated but an unbiased estimator for it is available as

$$v(e_{HR}) = e^2 - \left[\frac{N}{n}\sum_{i\in s}y_i^2 + \frac{N(N-1)}{n(n-1)}\sum\sum_{i\neq j\in s}y_i y_j\right] \tag{4.1.10}$$

This often turns out negative and $v(e_{HR})$ then does not exist.

Another alternative we consider is θ as Y and its unbiased estimator is taken as the Horvitz-Thompson (HT) estimator (HTE) which is $e_{HT} = \sum_{i\in s}\frac{y_i}{\pi_i}$ based on the sample s of size n taken from the population by SRSWR in n draws with

$$\pi_i = 1 - \left(\frac{N-1}{N}^n\right) \text{ and } \pi_{ij} = 1 - 2\left(\frac{N-1}{N}^n\right) + \left(\frac{N-2}{N}^n\right).$$

Though

$$V(e_{HT}) = \sum_{i<j}^{N}\sum^{N}(\pi_i\pi_j - \pi_{ij})\left(\frac{y_i}{\pi_i} - \frac{y_j}{\pi_j}\right)^2 + \sum_{1}^{N}\frac{y_i^2}{\pi_i}\alpha_i \text{ with}$$

$$\alpha_i = 1 + \frac{1}{\pi_i}\sum_{\substack{j=1\\j\neq i}}^{N}\pi_{ij} - \sum_{1}^{N}\pi_i,$$

to work out an appropriate sample size n it is needful to consider an unbiased estimator for $V(e_{HT})$, namely,

$$v(e_{HT}) = \sum\sum_{i<j\in s}\frac{(\pi_i\pi_j - \pi_{ij})}{\pi_{ij}}\left(\frac{y_i}{\pi_i} - \frac{y_j}{\pi_j}\right)^2 + \sum_{i\in s}\frac{y_i^2\alpha_i}{\pi_i^2} \tag{4.1.11}$$

In choosing the sample size fairly using Chebyshev's inequality we shall adopt the following procedure given by Chaudhuri and Dutta (2018) recalling the stipulation $100f = \frac{\mathrm{CV}(e)}{\sqrt{\alpha}}$ writing $\mathrm{CV}(e) = 100\frac{\sqrt{\widehat{V}(e)}}{e}$, $\widehat{V}(e) =$ an unbiased estimator of $V(e)$. Chaudhuri and Dutta (2018) enjoined the steps:

Step 1.   Obtain an unbiased estimator $v(e)$ based on an arbitrarily chosen size n for the sample according to a design p.

Step 2.   Repeat the Step 1 getting an estimated coefficient of variation $\mathrm{CV}(e) = 100\frac{\sqrt{v(e)}}{e}$

Step 3.   Compare the realized values of $\mathrm{CV}(e)$ versus the quantity $T = 100f\sqrt{\alpha}$ and take the right choice of n as n_0 for which $\mathrm{CV}(e)$ comes closest to T and rounded up to an integer reached on rounding off.

To fix the sample size in SRSWOR in estimating population total by Hartley-Ross estimator rather than by the expansion estimator this approach by Chaudhuri and Dutta (2018) is appropriate, provided $v(e_{HR}) > 0$.

To fix the number of draws in SRSWR while estimating the population total by the Horvitz and Thompson (1952) estimator also this Chaudhuri and Dutta (2018) procedure seems worth trying again starting with a preliminary n determined while estimating the population mean by the sample mean and then adjusting in successive steps, of course rounding upwards to the closest integer.

For calculations one may consult (Chaudhuri and Dutta, 2018; Chaudhuri, 2020 & Chaudhuri and Samaddar, 2022).

4.2   Controlled Sampling

So far we were proceeding with the presumption that once we identify the individual elements in a finite population it is a simple matter to choose a probability sample to render the survey data amenable to statistical analysis by assigning selection probability individually to them and implementing the selection easily consulting a table of random numbers. But often we encounter hazardous situations and face many hassles while in a field to choose from the population of sampling units. We may face difficult terrains in geographical realities. A population may contain hostile and ferocious human beings, insurmountable hillocks, forest areas, distant locations far from the main hospitable localities, racial composition of a population of interest may be so heterogeneous that field workers may not all be welcome to the host sample entities. In certain such situations the sampling units may be conveniently to be supposed to be composed of at least of two types, namely (1) some are non-preferred and (2) the others are 'preferred' units. Then it may be judicious to keep them separate and assign nil or negligible selection-probabilities to each of the 'non-preferred' units which together may be assigned a limited magnitude of selection probability, say α (0.05 or 0.01) and the higher probability $(1 - \alpha)$ may be distributed among the complementary set of the 'preferred' units.

A theoretically sound concept of 'Controlled sample selection' procedure to tackle such situations was introduced by Goodman and Kish (1950). This was further developed by Sukhatme and Avadhani (1965) and more developments followed through the efforts by Avadhani and Sukhatme (1965, 1966, 1968, 1972, 1973). Further refinements followed suite, vide (Wynn, 1977; Foody & Hedayat, 1977; Sinha, 1976; Sengupta, 1979, 1982; Gupta et al., 1982, 2012; Nigam et al., 1984, and most importantly Rao and Nigam (1990, 1992) have enormously contributed to the growth of this subject.

However, here we shall cover this topic very briefly. Suppose we contemplate taking from a population $U = (1, \ldots, i, \ldots, N)$ of size N an SRSWOR of $n(0 \le n < N)$ units and employ for the population mean $\bar{Y} = \frac{1}{N} \sum_1^N y_i$ the sample mean $\bar{y}$ which has the variance $\left(\frac{1}{n} - \frac{1}{N}\right) S^2$ where $S^2 = \frac{1}{N-1} \sum_1^N \left(y_i - \bar{Y}\right)^2$ and $\bar{y}$ is unbiased for $\bar{Y}$.

In this case the inclusion-probabilities are

$$\pi_i = \frac{n}{N} \forall i \in U \text{ and } \pi_{ij} = \frac{n(n-1)}{N(N-1)} \forall i, j \in U (i \ne j).$$

A controlled sampling demands instead of the SRSWOR a different sampling scheme avoiding selection of non-preferred units but confining selection to a smaller set of sample units but retaining π_i and π_{ij} in tact as above so that $\bar{Y}$ may be unbiasedly estimated by $\bar{y}$ retaining the same variance $V(\bar{y})$ as by SRSWOR, ie, $V(\bar{y}) = \left(\frac{1}{n} - \frac{1}{N}\right) S^2$. It is noteworthy that SRSWOR has the support size as $\binom{N}{n}$, i.e., it assigns selection probability positively to $\binom{N}{n}$ possible samples but the above-noted alternative sampling design has the support size as $\binom{N}{n} - \binom{N}{2}$ because every sample with positive selection probability should be such that the $\binom{N}{2}$ conditions that $\pi_{ij} = \frac{n(n-1)}{N(N-1)} \forall i, j \in U (i \ne j)$ must be satisfied. So, an immediate impact of controlled sampling is that the support size is reduced in discarding 'non-preferred' samples.

Chaudhuri and Stenger (2005) give a simple illustration. Suppose $N = 9$ and an SRSWOR of n units, $n = 3$ is contemplated to be chosen to yield $\bar{y}$ as an unbiased estimator for $\bar{Y}$. Here each of the $\binom{9}{3}$ samples is given a selection probability $\frac{1}{\binom{9}{3}}$ and $\pi_i = \frac{1}{3} \forall i \in U = (1, \ldots, 9)$ and $\pi_{ij} = \frac{1}{12} \forall i, j \in U = (1, \ldots, 9)$ but $i \ne j$.

But suppose only the following 12 samples are identified as $(1, 2, 3)$, $(1, 6, 8)$, $(2, 4, 9)$, $(3, 5, 7)$, $(1, 5, 9)$, $(1, 4, 7)$, $(2, 5, 8)$, $(2, 6, 7)$, $(3, 4, 8)$, $(3, 6, 9)$, $(4, 5, 6)$ and $(7, 8, 9)$ are identified as the 'preferred' samples. Thus, instead of the SRSWOR with $p(s) = \frac{1}{\binom{9}{3}}$ for every sample s out of the $\binom{9}{3}$ samples one may rather adopt the sampling design q which to each of the above 12 samples assigns the probability of selection as $q(s) = \frac{1}{12}$ to each of the above illustrated 12 samples only. Then, for this design q also as for the SRSWOR p above one may check that (i) $\pi_i = \frac{1}{3}$ for every $i = 1, \ldots, 9$ and (ii) $\pi_{ij} = \frac{1}{12}$ for every $i, j = 1, \ldots, 9 (i \ne 5)$.

So, the sample mean $\bar{y}$ based on q is unbiased for $\bar{Y}$ and has the same variance $V(\bar{y})$ equal to $\left(\frac{1}{3} - \frac{1}{9}\right) S^2$, where

$$S^2 = \frac{1}{8} \sum_1^9 \left(y_i - \bar{Y}\right)^2, \bar{Y} = \frac{1}{9} \sum_1^9 y_i.$$

This design q is an example of a 'controlled design'. Besides the above 12 samples all other $\binom{9}{3} - 12 = 84 - 12 = 72$ samples are assigned zero selection probabilities though π_i and π_{ij}'s are each positive for $i = 1, \ldots, 9$ and $i, j = (1, 2), (1, 3), \ldots, (8, 9)$ for the controlled design q. But in the present case numerous alternative controlled designs may be constructed. As against the SRSWOR with N and n and the support size $\binom{N}{n}$ a controlled design ensuring the same expectation $\bar{Y}$ and $\left(\frac{1}{n} - \frac{1}{N}\right) S^2$, for the variance of the sample mean can be constructed on ensuring $\pi{ij} = \frac{n(n-1)}{N(N-1)}$ (which immediately implies $\pi_i = \frac{n}{N} \forall i$) $\forall i, j$ in $U = (1, \ldots, N)(i \neq j)$. So, with a support size $\binom{N}{n} - \binom{N}{2}$ innumerable controlled designs may be constructed. By Support we mean the set of units with positive selection probability.

A standard procedure to construct a controlled design corresponding to an SRSWOR (N, n) design is to follow Chakrabarti (1963) to get hold of a BIBD (Balanced Incomplete Block Design) with N treatments, b blocks of size n each such that each treatment occurs altogether a number of times r and every pair of treatments occurs together in a block a number of times λ with the restriction (I) $bk = Nr$ and (II) $\lambda = \frac{bn(n-1)}{N(N-1)}$, a treatment here is a unit and select at random (i.e., with probability $\frac{1}{b}$) one of these b blocks, to get a controlled design.

It then follows that $\pi_{ij} = \frac{\lambda}{b} = \frac{n(n-1)}{N(N-1)} \forall i, j \in U = (1, \ldots, N)$ and $\pi_i = \frac{n}{N} \forall i \in U$. Here the support size for this controlled design is b. But to ensure a required control it is important that one needs $b < \binom{N}{n}$. Unfortunately, sometimes, say, when $N = 8$ and $n = 3$ a BIBD does not exist. Many controlled designs have been constructed in the literature devising ways to circumvent this inadequacy.

Gupta et al. (1982) used BIBD to work out controlled varying probability sampling designs reducing the support size vis-a-vis an original sampling scheme using normed positive size-measures $p_i (0 < p_i < 1)$, sample size n, the inclusion-probabilities $\pi_i = np_i$ and π_{ij} and employing the Horvitz-Thompson (HT, 1952) estimator HTE, namely

$$t_{HT} = \sum_{i \in b} \frac{y_i}{\pi_i} \text{ with variance}$$

$$V(t_{HT}) = \sum_{\substack{i=1 \\ i \leq j}}^{N} \sum_{j=1}^{N} (\pi_i \pi_j - \pi_{ij}) \left(\frac{y_i}{\pi_i} - \frac{y_j}{\pi_j}\right)^2;$$

when every sample has a fixed number of distinct units n.

Nigam et al. (1984) also pursued with the same basic problem. Gupta et al. (2012) is the most comprehensive source of materials on this topic. But whether the research activities by Sinha (1976), Sengupta (1979) and others in this context or by the early pioneers (Goodman & Kish, 1950) by procedures adopted by them is worth pursuing seems doubtful after perusing the decisive work on this topic by Rao and Nigam (1990, 1992).

They have spelled out that the 'Simplex Method' in the context of 'Linear Programming', vide (Taha, 1976; Dantzig, 1963) provides an easy and comprehensive solution to this 'Controlled Sampling' problem.

If an original sampling design p with support S such that

$$p(s) \geq 0 \,\forall s \text{ in } S, \sum_{s \ni i,j} p(s) = \pi_{ij}$$

$$\text{and} \sum_{\substack{j=1 \\ \neq i}} \pi_{ij} = (n-1)\pi_i = n(n-1)p_i$$

$$(0 < p_i < 1 \,\forall i \in U = (1, \ldots, i, \ldots, N))$$

is contemplated to be initially employed with estimator $t_{HT} = \sum_{i \in s} \frac{y_i}{\pi_i}$ having the variance

$$V(t_{HT}) = \sum_{i<j}^{N} \sum^{N} (\pi_i \pi_j - \pi ij) \left(\frac{y_i}{\pi_i} - \frac{y_j}{\pi_j} \right)^2,$$

then suppose the support has to be reduced to a subset of S and a controlled design has to be devised. Then, Rao and Nigam (1990, 1992) recommend resorting to the Simplex Method so as to minimize objective function

$$\phi = \sum_{s \in S_1} p(s)$$

taking and identifying S_1 as the subset of 'Non-preferred' samples in S subject to p meeting the requirements specified above with numerical values of π_i, π_{ij}, n, N specified along with $p_i, i = 1, \ldots, N$ so that the $V(t_{HT})$ above be attained for the y_i's, $i = 1, \ldots, N$.

A few more additional but not compelling ramifications are also indicated in Rao and Nigam (1990, 1992).

4.3 Lessons and Exercises

Lessons

(i) Consider the model

$$y_i = \alpha + \beta x_i + \epsilon_i, i = 1, \ldots, N$$

Here N, α, β are assignable constants; x_i's $(i = 1, \ldots, N)$ are random variables independently distributed exponentially as

$$f(x) = e^{-x}, 0 < x < +\infty$$
$$\text{and } F(x) = \text{Prob}(X \leq x) = 1 - e^{-x}, 0 < x < \infty;$$

ϵ_i's are independently normally distributed identically with zero mean and variance unity.

Take N as 125, $\alpha = 0.2$, $\beta = 3.4$ and generate y_i's for $i = 1, \ldots, i, \ldots, N$. For this take random numbers reduced to fractions correct to 4 places after decimal equal to $F(x)$ above.

Present your generation in a table. You may refer to this table many times later while going through this book.

(ii) Utilize the table referred to in (i) above, tabulate an exercise to show how you choose suitably a number of units in a sample to be drawn by SRSWR so as to unbiasedly estimate the population mean by the sample mean of the distinct units in the sample, showing the performance of the over-all sample mean from a sample in 12 draws as a reference illustration.

Exercises

(i) Utilizing the table referred to in the model of lesson above on taking SRSWOR's of sizes 4, 9, 12 and 15 show how the Hartley-Ross estimator for the population total of y_i's may fare as a good rival against the sample mean with a well-chosen sample size.

(ii) Referring to the Table above take the first 20 units as a sub-population. From this take an SRSWOR of 12 units. Keeping this sample aside, mark off 4 of its units as 'non-preferred', take a controlled sample. Taking these 12 units as a sub-population describe how you may take from it a "Controlled equal probability sample" of size 3.

References

Chaudhuri, A. (2010). *Essentials of survey sampling*. New Delhi, India: PHI.

Hartley, H. O., & Ross, A. (1954). Unbiased ratio estimators. *Nature, 174,* 270–271.

Chaudhuri, A., & Dutta, T. (2018). Determining the size of a sample to take from a finite population. *Statistics and Applications, 16*(1), 37–44 New series.

Horvitz, D. C., & Thompson, D. J. (1952). A generalization of sampling without replacement from a finite universe. *JASA, 47,* 663–689.

Chaudhuri, A. (2020). A review on issues of settling the sample-size in surveys: Two approaches: Equal and varying probability sampling—crises in sensitive issues. *CSA Bulletin, 72*(1), 7–16.

Chaudhuri, A., & Samaddar, S. (2022). Estimating gains in efficiency of complex over simple sampling strategies and studying relative model variance of pair-wise rival strategies with simulated illustrations. Statistics in Transition, New Series, Accepted on 31 March, 2021 for publication in 2022.

Goodman, L. A., & Kish, L. (1950). Controlled selection'A technique in probability sampling. *JASA, 55,* 350–372.

Sukhatme, B. V., & Avadhani, M. S. (1965). Controlled selection: a technique random sampling. *AISM, 17,* 15–28.

Avadhani, M. S., & Sukhatme, B. V. (1965). Controlled random sampling. *Journal of the Indian Society of Agricultural Statistics, 17*(1), 34–42.

Avadhani, M. S., & Sukhatme, B. V. (1966). A note on ratio and regression methods of estimation under controlled simple random sampling. *Journal of the Indian Society of Agricultural Statistics, 18*(2), 17–20.

Avadhani, M. S., & Sukhatme, B. V. (1968). Simplified procedures for designing controlled simple random sampling. *Australian Journal of Statistics, 10,* 1–7.

Avadhani, M. S., & Sukhatme, B. V. (1972). Sampling on several successive occasions with equal and unequal probabilities and without replacement. *Australian Journal of Statistics, 14,* 109–119.

Avadhani, M. S., & Sukhatme, B. V. (1973). Controlled sampling with equal probabilities and without replacement. *International Statistical Review, 41,* 175–182.

Wynn, H. P. (1977). Convex sets of finite population plans. *Annals of Statistics, 5,* 414–418.

Foody, W., & Hedayat, A. (1977). On theory and applicability of BIB designs with repeated blocks. *The Annals of Statistics, 5,* 932–945.

Sinha, B. K. (1976). On balanced sampling scheme. *CSA Bulletin, 25,* 129–138.

Sengupta, S. (1979). On the construction of non-invariant balanced sampling designs. *CSA Bulletin, 28,* 109–124.

Sengupta, S. (1982). Construction of some non-invariant balanced sampling designs. *CSA Bulletin, 31,* 165–184.

Gupta, V. K., Nigam, A. K., & Kumar, P. (1982). On a family of sampling schemes with inclusion probability proportional to size. *Biometrika, 69,* 191–196.

Nigam, A. K., Kumar, P., & Gupta, V. K. (1984). Some methods of inclusion probability proportional to size sampling. *JRSS, 46,* 564–571.

Gupta, V. K., Mandal, B. N., & Parsad, R. (2012). *Combinatorics in sample surveys vis-a-vis controlled selection.* Searbrucken, Germany: Lambert Academic Publ.

Rao, J. N. K., & Nigam, A. K. (1990). Optimal controlled sampling designs. *Biometrika, 77,* 807–814.

Rao, J. N. K., & Nigam, A. K. (1992). Optimal controlled sampling: A unified approach. *International Statistical Review, 60,* 89–98.

Chaudhuri, A., & Stenger, H. (1992, 2005). *Survey sampling theory and methods* (1st ed.). Marcel Dekker, N.Y.; (2nd ed.) CRC Press, Florida, N.Y., USA.

Chakrabarti, M. C. (1963). On the use of incidence matrices in sampling from finite populations. *JISA, 1,* 78–85.

Taha, H. A. (1976). *Operations research: An introduction* (2nd ed.). NY, USA: McMillan Inc.

Dantzig, G. B. (1963). *Linear programming and extensions.* Santa Monica, CA: RAND corp.

Chapter 5
Adjusting Unit-Nonresponse by Weighting and Tackling Item-Nonresponse by Imputation

5.1 Weight Adjusting Methods of Unit-Nonresponse

We shall illustrate now several weight adjusting methods to take care of unit non-responses.

5.1.1 Method of "Call backs"

Suppose an SRSWOR of $n(2 \leq n < N)$ units is chosen from population of size N. But suppose required data are gathered from a sub-sample of $n_1(2 \leq n_1 < n)$ sampled units but $n_2 = n - n_1$ units do not respond at all. To develop a theory it is assumed that the population of N units consists of an unknown number of $N_1(< N)$ units who will yield responses on being approached as are the sampled n units but the remaining $N_2 = N - N_1$ units will remain non-responding units. But about these N_2 initial non-respondents it is assumed that with an improved data gathering device being employed a second sample taken out of the n_2 initial non-respondents belonging to these N_2 individuals may agree to give out the desired responses. Now suppose a certain number r out of these n_2 initial non-respondents are chosen by SRSWOR and they agree to give out the data demanded of them.

Had all the n sampled units responded we could obtain for $\bar{Y} = \frac{1}{N} \sum_1^N y_i$ the unbiased estimator

$$\bar{y} = \frac{1}{n} \sum_{i \in s} y_i$$

calling s the initial sample of size n.

© The Author(s), under exclusive license to Springer Nature Singapore Pte Ltd. 2022
A. Chaudhuri and S. Pal, *A Comprehensive Textbook on Sample Surveys*, Indian Statistical Institute Series, https://doi.org/10.1007/978-981-19-1418-8_5

Now $\sum_{i \in s} y_i = n_1 \bar{y}_1 + n_2 \bar{y}_2$

writing $\bar{y}_1 = \dfrac{1}{n_1} \sum_{i \in s_1} y_i,\ s_1 =$ initial respondents

and $\bar{y}_2 = \dfrac{1}{n_2} \sum_{i \in s_2} y_i,\ s_2 =$ initial non-respondents,

since in $\bar{y} = \dfrac{1}{n}(n_1 \bar{y}_1 + n_2 \bar{y}_2)$

the quantity $\bar{y}_2$ is unascertainable, let from s_2 of n_2 non-respondents an SRSWOR s_r of size $r(2 \le r < n_2)$ be chosen and surveyed obtaining an unbiased estimator

$$\bar{y}_r = \frac{1}{r} \sum_{s_r} y_i \text{ for } \bar{y}_2 \text{ be obtained.}$$

$$\text{Then, } \bar{y}' = \frac{1}{n}(n_1 \bar{y}_1 + n_2 \bar{y}_r) = \frac{n_1}{n} \bar{y}_1 + \frac{n_2}{n} \bar{y}_r$$

may be taken as an unbiased estimator for $\bar{y}$ by virtue of the method of choosing s_r.

This method of choosing an initial sample s and a second sample s_r from s_2 is called Two-phase or Double sampling. So, by E_1, E_2 and $E = E_1 E_2$ the 1st phase, 2nd phase and over-all expectation operators and by $V_1, V_2, V = E_1 V_2 + V_1 E_2$ the 1st phase, 2nd phase and over-all variance-operators are denoted.

$$\text{Then, } E_2 \bar{y}_r = \bar{y}_2,\ E_2(\bar{y}') = \frac{1}{n}(n_1 \bar{y}_1 + n_2 \bar{y}_2) = \bar{y}$$

and $E(\bar{y}') = E_1(\bar{y}) = \bar{Y}$;

also, $V_2(\bar{y}_r) = \left(\dfrac{1}{r} - \dfrac{1}{n_2}\right) \dfrac{1}{(n_2 - 1)} \sum_{s_2} (y_i - \bar{y}_2)^2$

$$E_1 V_2(\bar{y}_r) = \left(\frac{1}{r} - \frac{1}{n_2}\right) \frac{1}{(N_2 - 1)} \sum_{1}^{N_2} (y_i - \bar{y}_2)^2$$

writing $\bar{y}_2 = \dfrac{1}{N_2} \sum_{i=1}^{N_2} y_i,\ \dfrac{1}{N_2 - 1} \sum_{1}^{N_2} (y_i - \bar{y}_2)^2 = S_2^2$

or $E_1 V_2(\bar{y}_r) = \dfrac{1}{n_2}(K - 1)S_2^2$, writing $K = \dfrac{n_2}{r}$.

$$\text{So, } V(\bar{y}') = V_1 E_2(\bar{y}') + E_1 V_2(\bar{y}')$$

$$= V_1(\bar{y}) + E_1 \left(\frac{n_2}{n}\right)^2 \left(\frac{K-1}{n_2}\right) S_2^2$$

$$= \left(\frac{1}{n} - \frac{1}{N}\right) S^2 + \left(\frac{K-1}{n_2}\right) W_2 S_2^2$$

writing $S^2 = \frac{1}{N-1} \sum_1^N (y_i - \bar{y})^2$ and $W_2 = \frac{N_2}{N}$ because $E_1 \left(\frac{n_2}{n}\right) = W_2$ since (n_1, n_2)

has the hypergeometric distribution with $E_1 \left(\frac{n_1}{n}\right) = \frac{N_1}{N} = W_1$, say and $E_1 \left(\frac{n_2}{n}\right) = \frac{N_2}{N} = W_2$.

Thus, if there was a 100% response $\bar{y} = \frac{1}{n} \sum_{i \in s} y_i$ is an unbiased estimator for $\bar{Y}$

with $\frac{1}{n}$ as the uniform weight for each observed unit.

But on encountering n_1 responses initially and r responses out of the initial n_2 non-responses, a corresponding unbiased estimator $\bar{y}' = \frac{n_1}{n} \sum_{i \in s_1} y_i + \frac{n_2}{nr} \sum_{i \in s_r} y_i$ so that

initial respondents have the weight $\frac{1}{n}$ and $\frac{n_2}{nr}$ as the weight for every initial non-respondent who is a respondent on the second sampling effort, every such responding having a higher weight.

This is known as the "Call-back" method of encountering the unit non-response problem because a first-time non-respondent is paid a second visit to extract a response.

The mathematical simplicity of the variance formula

$$V(\bar{y}') = \left(\frac{1}{n} - \frac{1}{N}\right) S^2 + \frac{(K-1)}{n} W_2 S_2^2 = V' \tag{5.1.1}$$

may be utilized as follows in working out rules for suitable choice of n and r.

In contrast with the 'variance function' (5.1.1) involving the assignable variables n and K and the constants N, S^2 and S_2^2 it is advisable to consider 'Cost function'

$$C' = nC_0 + n_1 C_1 + r C_2 \tag{5.1.2}$$

involving an over-all cost in surveying each of the initial sample if size n, another cost component C_1 (obviously $C_1 > C_0$) per unit among the initial respondents n_1 in number and a third cost component C_2 per unit in the second phase sample of size r (obviously $C_2 > C_1$).

The formula (5.1.1) yields

$$\left(V' + \frac{S^2}{N}\right) = \frac{1}{n} \left(S^2 - W_2 S_2^2\right) + \frac{1}{n} \left(K W_2 S_2^2\right) = V, \text{ say.}$$

Conveniently we may write (5.1.2) as

$$C' = n\left(C_0 + C_1\frac{n_1}{n} + C_2\frac{n_2}{n}\frac{1}{K}\right) \text{ and so}$$

$$E(C') = n\left(C_0 + C_1 W_1 + C_2\frac{W_2}{K}\right) = C, \text{ say.}$$

This yields $VC = \alpha + \beta K + \dfrac{\theta}{K}$ \hfill (5.1.3)

on writing $\alpha = \left(S^2 - W_2 S_2^2\right)(C_0 + C_1 W_1)$
$$\beta = (C_0 + C_1 W_1)\, W_2 S_2^2 \text{ and}$$
$$\theta = C_2 W_2\left(S^2 - W_2 S_2^2\right)$$

The product VC involves a single variable K. Then it is advisable to minimize $\alpha + \beta K + \frac{\theta}{K}$ with respect to K which will enable us to justify an action either to minimize V for a fixed C or alternately minimize C for a fixed V. To choose such a K we may write

$$\beta K + \frac{\theta}{K} \text{ as } \left(\sqrt{\beta K}\right)^2 + \left(\sqrt{\frac{\theta}{K}}\right)^2$$

$$= \left(\sqrt{\beta K} - \sqrt{\frac{\theta}{K}}\right)^2 + 2\sqrt{\beta\theta}$$

which will be minimized for the choice

$$K_{\text{opt}} = \sqrt{\frac{\theta}{\beta}} \text{ for } K.$$

If β and θ were known of course we should take r as $\frac{n_2}{K_{\text{opt}}}$.

If V were settled an optimal choice of n would be

$$n_{\text{opt}} = \left(S^2 - W_2 S_2^2 + K_{\text{opt}} W_2 S_2^2\right)/V.$$

Alternatively, if the cost C be fixed, then an optimal choice of n would be

$$n_{\text{opt}} = C/\left[C_0 + C_1 W_1 + C_2\frac{W_2}{K_{\text{opt}}}\right].$$

This theory is due to Hansen and Hurwitz (1946).

Choosing K_{opt} and n_{opt} is not practicable because they involve unknowable parameters. They may be estimated from Pilot Surveys—a topic we have not described yet but postponed to the final chapter of this volume.

5.1.2 Disposing of the 'Not-at-Homes'

Politz and Simmons (1949) coined this concept encountering the situations when in human surveys of households knocking at the doors no one was found to answer the calls. They recommend classifying those, who are found, to be classified into 8 mutually exclusive categories $h = 0, 1, 2, \ldots, 7$ namely those who reported that they were at home to answer the calls on the h of the preceding days ($h = 0, 1, 2, \ldots, 7$). Their method of analysis started on assigning the weight $\left(\frac{8}{h+1}\right)$ to the value y_i for the i-th responding unit assigned to the h-th category during the survey.

Thus while employing the Horvitz and Thompson (1952) method to estimate the population total $Y = \sum_1^N y_i$, namely $\widehat{Y}_{HT} = \sum_{i \in s} \frac{y_i}{\pi_i}$, the Politz-Simmons's prescription is to revise this into

$$\widehat{Y}_{HT}(PS) = \frac{\left[\sum_{h=0}^7 \left(\sum_{i \in s} \frac{y_{ih}}{\pi_i} \left(\frac{8}{h+1}\right)\right)\right]}{\sum_0^7 W_h} = \frac{\sum_{h=0}^7 W_h \sum_{i \in s} \frac{y_{ih}}{\pi_i}}{\sum_{h=0}^7 W_h},$$

$$\text{(writing } W_h = \frac{8}{h+1}.$$

Here the weight assigned to a sample person or sample household labelled i assigned to the hth category the weight $\left(\frac{8}{h+1} \frac{1}{\pi_i}\right), i \in s$ if ith unit is found to be in the hth category. This clearly is a case of weighting adjustment.

Though the weights $W_h = \frac{8}{h+1}$ are determined from sampled units if we treat them as constants ignoring the errors involved then writing

$$\widehat{Y}_{HT}(PS) = \sum_{h=0}^7 Q_h \sum_{i \in s} \frac{y_{ih}}{\pi_i}$$

and $Q_h = \frac{W_h}{\sum_0^7 W_h}$, treated as constants we may approximately take

$$E\left(\widehat{Y}_{HT}(PS)\right) = \sum_{h=0}^7 Q_h Y_h = Y$$

with

$$Y_h = \sum_{i=1}^N y_{ih}.$$

Thus $\widehat{Y}_{HT}(PS)$ is approximately unbiased for Y.

Also, the approximate variance of $\widehat{Y}_{HT}(PS)$ is

$$V\left(\widehat{Y}_{HT}(PS)\right) = \sum_{h=0}^{7} Q_h^2 \left[\sum_{i<j}^{N} \sum^{N} \left(\pi_i \pi_j - \pi_{ij}\right) \left(\frac{y_{ih}}{\pi_i} - \frac{y_{jh}}{\pi_j}\right)^2 \right]$$

and an approximately unbiased estimator for it is

$$v\left(\widehat{Y}_{HT}(PS)\right) = \sum_{h=0}^{7} \widehat{Q}_h^2 \sum_{i<j\in s} \sum \frac{\left(\pi_i \pi_j - \pi_{ij}\right)}{\pi_{ij}} \left(\frac{y_{ih}}{\pi_i} - \frac{y_{jh}}{\pi_j}\right)^2$$

provided $\pi_{ij} > 0 \,\forall\, i \neq j$ in $U = (1, \ldots, i, \ldots, N)$.

If one contemplates employing any other estimator for Y choosing out of a large number of such ones illustrated so far in this text like (Rao et al., 1962) or (Lahiri, 1951) unbiased ratio estimator, then also similarly as above Politz and Simmons (1949) method to tackle unit non-response may be applied similarly with little difficulty.

5.1.3   A Model-Based Weighting Approach to Counter Unit Non-response

Let $I_{ri} = 1$ if ith unit responds on request

$\quad\quad = 0$, else.

Let I_{ri}'s be independent random variables with expectations and variances

$$E\left(I_{ri}\right) = q_i \left(0 < q_i < 1 \,\forall\, i\right) \text{ and}$$
$$V\left(I_{ri}\right) = \psi_i \left(0 < \psi_i \,\forall\, i\right)$$

$$\text{Let } \log_e\left(\frac{q_i}{1 - q_i}\right) = \alpha + \beta z_i \quad\quad\quad (5.1.4)$$

with α, β as unknown constants and $z_i = (i = 1, \ldots, N)$ as ascertainable numbers at the investigator's hands. Let in terms of z_i's $(i \in U)$ one may obtain estimates q_i's $(0 < \widehat{q}_i) < 1 \,\forall\, i)$ for q_i be available subject to

$$\log_e\left(\frac{\widehat{q}_i}{1 - \widehat{q}_i}\right) = \alpha + \beta z_i + e_i, i \in s$$

with e_i's as unobservable random variables independently distributed over i in U.

Let α, β be estimated by least squares which yield the estimates $\widehat{q}_i$, $i \in s$ because (5.1.4) gives

$$q_i = \frac{e^{\alpha+\beta z_i}}{1 + e^{\alpha+\beta z_i}} = \frac{1}{1 + \bar{e}^{(\alpha+\beta z_i)}}, i \in U,$$

the 2-parameter logistic regression model.

Now, if the (Horvitz & Thompson, 1952) estimator be contemplated to be used to estimate $Y = \sum_1^N y_i$, then if some sampled units be anticipated not to respond, then the estimator

$$E_{HT} = \sum_{i \in s} \frac{y_i}{\pi_i}, \text{ then.}$$

this is to be revised by the estimator

$$t_{HT}(NR) = \sum_{i \in s} \frac{y_i}{\pi_i} \frac{I_{r_i}}{\widehat{q}_i}$$

because $E_R(t_{HT}(NR)) = \sum_{i \in s} \frac{y_i}{\pi_i} \frac{E_R(I_{r_i})}{q_i} = t_{HT}$ neglecting the error

$$E_R\left(\frac{1}{\widehat{q}_i}\right) - \frac{1}{q_i}, i \in s.$$

By E_R we mean the expectation in respect of the response variables I_{r_i}, $i \in U$.

$$V(t_{HT}(NR)) = \sum_{i<j}^N \sum^N (\pi_i \pi_j - \pi_{ij}) \left(\frac{y_i I_{r_i}}{\pi_i q_i} - \frac{y_j I_{r_j}}{\pi_j q_j}\right)^2$$

and an approximately unbiased estimator for it is

$$v(t_{HT}(NR)) = \sum_{i<j\in s} \sum \left(\frac{\pi_i \pi_j - \pi_{ij}}{\pi_{ij}}\right) \left(\frac{y_i I_{r_i}}{\pi_i \widehat{q}_i} - \frac{y_j I_{r_j}}{\pi_j \widehat{q}_j}\right)^2$$

assuming $\pi_{ij} > 0 \forall i \neq j \in U$.

If instead of Horvitz-Thompson estimator some other estimator is contemplated to be used then also similar weight-adjusted estimator as above may be similarly employed.

More details about how to encounter unit non-responses one should consult (Cochran, 1977; Chaudhuri and Stenger, 1992; Chaudhuri, 2014).

5.2   Imputation Techniques to Control Item Non-responses

In almost every large-scale survey where a population is composed of a number of strata and on it a survey is carried out at multiple stages and in more than one phases in respect of every ultimate unit numerous items of enquiry are identified. But across the units the numbers of items on which intended data are recorded vary in enormous magnitudes. Since computer-based data-analysis is the order of the day, technically it is undesirable to record enormously varying numbers of item-wise entries because of item non-responses. A technically useful device being employed is to 'put in some value' for each missing item on varying numbers of units. This is called 'Imputation technique'. But the 'values imputed' on the missing units must not be arbitrarily assigned so as to avoid wild departures from truth. As a matter of fact some justifiable procedure must be followed to prescribe a value to take the place of a missing one for a unit. Hence have appeared diverse procedures for inducing imputations.

5.2.1   Logical, Consistency or Deductive Imputation

For certain items codes are used and from the nature of related items filled-in item code for a missing item like sex, age-group, marital status, income code or educational code may be correctly imputed as a matter of logical consequence or from the consistency point of view.

5.2.2   Cold Deck Imputation

Certain missing values may be retrieved from past records or from survey records on an allied subject in the contemporary period. These are called 'Cold Decks' from which appropriate values may be quoted to be imputed for the relevant current one.

5.2.3   Mean Value Imputation

In the context of Imputation one needs to specify certain Imputation Classes which denote certain Population units bracketed into certain groups with common ranges of values or codes in respect of certain related items.

If for a unit an item value is missing it may be replaced by the mean of the realized values for the units in a common Imputation class. This is of course an easy procedure but is not statistically worthy because while using mean-values, the variability turns out suppressed and concealed which is statistically nothing but detrimental.

5.2.4 Hot Deck Imputation

As opposed to Cold Deck Imputation, the Hot Deck Imputation technique substitutes values for missing unit from those on record in the current survey itself.

Suppose the ultimate units in the survey are labelled as

$$i_1, i_2, i_3, i_4, i_5, i_6, i_7, i_8, i_9, i_{10}, i_{11}$$

and on a variable y of interest values of a variable from the concerned current survey are only y_{i_3}, y_{i_6} and y_{i_9} and for the remaining units y-values are all missing. Now suppose the concerned y-values relate to an 'Imputation class' from a cold deck a related x-value is available as, say, x. Then, let us denote the imputed values as $z_{i_1}, z_{i_2}, \ldots, z_{i_{11}}$. Then these are taken as $z_{i_1} = z_{i_2} = x, z_{i_3} = y_{i_3} = z_{i_4} = z_{i_5}, z_{i_6} = y_{i_6} = z_{i_7} = y_{i_6} = z_{i_8}, z_{i_9} = y_{i_9}, z_{i_{10}} = y_{i_9} = z_{i_{11}}$.

The unit which has an intended value of interest and the value of which is substituted for a unit with a missing value is called a 'Donor' and the unit for which an intended value is 'missing' but the missing value is substituted by the value of some unit other than itself is called a 'Recipient' in the context of 'Imputation'. In the present example, the units i_3, i_6 and i_9 are the 'Donors' and the units $i_4, i_5, i_7, i_8, i_{10}$ and i_{11} are the 'Recipients'.

5.2.5 Random Imputation

Suppose from the units of the hth Imputation class a sample of n_h units have been taken and out of them $r_h(1 < r_h < n_h)$ units 'respond' and $m_h = n_h - r_h$ units 'do not respond' on the item of interest. Let two numbers t_h and K_h be fixed up for every $h(= 1, \ldots, H)$, say. The rate of response over-all, namely, $\frac{r}{n}$, writing $r = \sum_{h=1}^{H} r_h$

and $m = \sum_{h=1}^{H} m_h$, must not be less than 50% though for respective h's, m_h may equal, exceed or fall short of $r_h(h = 1, \ldots, H)$. Now take an SRSWOR of size $t_h(1 < t_h < r_h)$ donors out of the r_h respondents and use each of their values a number $(K_h + 2)$ times and out of the $(r_h - t_h)$ remaining respondents, each of whom is also taken as donor, repeat each of their values $(K_h + 1)$ times.

Thus
$$n_h = t_h(K_h + 2) + (r_h - t_h)(K_h + 1)$$
$$= r_h + r_h K_h + t_h$$

implying
$$r_h K_h + t_h = m_h$$

which is the number of missing units out of the n_h units in the sample.

Taking $K_h = 0, 1, 2$ etc it is quite easy to have the simple rule of choosing donors and using and re-using their values for the 'missing' units. This is called 'Random Imputation' method.

5.2.6 Regression Imputation Technique

Suppose besides the variable of interest y there is another positively correlated variable x. Also in a sample taken in a survey the sample size is n and x_i values are observed for all the units $i, (i = 1, \ldots, n)$ but y_i-values are available on r of the responding units $(2 \le r < n)$ on the item of interest. Then for the postulated line of regression of y on x, namely

$$y = \alpha + \beta x + e$$

values available are denoted as, say,

$$y_i, i = 1, \ldots, r \text{ and } x_i, i = 1, \ldots, n.$$

Then the least square estimates of α, β are obtained as

$$a = \bar{y}(r) - b\bar{x}(r) \text{ and}$$

$$b = \frac{\sum_1^r (y_i - \bar{y}(r))(x_i - \bar{x}(r))}{\sum_1^r (x_i - \bar{x}(r))^2}$$

$$\text{writing } \bar{y}(r) = \frac{1}{r} \sum_1^r y_i \text{ and } \bar{x}(r) = \frac{1}{r} \sum_1^r x_i.$$

Then, the imputed values on y are $z_i = a + bx_i$ for $i = r + 1, \ldots, n$ by the Regression method of Imputation.

5.2.7 Multiple Imputation Methods

We have presented so far 5 methods of imputation.

Suppose as an estimator for a finite population parameter a statistic t_j has been obtained on applying the jth $(j = 1, 2, \ldots, C)$ method of imputation and v_j is the estimate of its variance $(j = 1, 2, \ldots, C)$. Then $\bar{t} = \frac{1}{C} \sum_{j=1}^{C} t_j$ is taken as the final estimator for θ and

$$\bar{v} = \frac{1}{C}\sum_{j=1}^{C} v_j + \frac{1}{(C-1)}\sum_{j=1}^{C}(t_j - \bar{t}^2)$$

as the over-all variance estimator for $\bar{t}$.

5.2.8 Repeated Replication Method Imputation

A selected sample is split up into disjoint parts numbering, say, C. For the respective parts one of the specific imputation method is applied, employing a different method for each of the other disjoint parts. Then, t_j with $j = 1, \ldots, C$ estimates for the parameter θ is obtained along with v_j as a serviceable variance estimator for t_j, $j = 1, \ldots, C$. Next as in Replicated sampling an estimator for θ the over-all estimator is taken as $\bar{t} = \frac{1}{C}\sum_{1}^{C} t_j$ and the over-all variance estimator is taken as

$$\bar{v} = \frac{1}{2C(C-1)}\sum_{j=1}^{C}\sum_{\substack{j'=1 \\ j \neq j'}}^{C}(t_j - t_{j'})^2.$$

For more details on Imputation one should consult (Rubin, 1986; 1988). These are of a much higher standards and so we do not enter into them to keep our level of coverage modest. Further theoretical and practical studies are of course necessary and rewarding.

Since imputation does not yield truth when it is amiss it is incumbent in a Survey Report to mention and describe palpably the extent and nature of imputations have been incorporated in various parts and aspects covered.

5.3 Lessons

(i) Clearly bring out the conceptual distinction between unit non-response and Item Non-response. Also distinguish between weighting adjustments and Imputation Techniques.
(ii) Explain the concepts of Imputation Classes, Donors and Recipients in the context of tackling Item Non-responses with illustrations.

References

Politz, A., & Simmons, W. (1949). Note on attempt to get the "not at homes" into the sample without call-backs. *JASA, 45,* 136–137.

Horvitz, D. G., & Thompson, D. J. (1952). A generalization of sampling without replacement from a finite universe. *JASA, 47,* 663–689.

Rao, J. N. K., Hartley, H. O., & Cochran, W. G. (1962). On a simple procedure of unequal probability sampling without replacement. *JRSS B, 24,* 482–491.

Lahiri, D. B. (1951). A method of sample selection providing unbiased ratio estimators. *BISI, 33*(2), 133–146.

Cochran, W. G. (1977). *Sampling techniques.* Delhi, India: Wiley Eastern Limited.

Chaudhuri, A., & Stenger, H. (1992, 2005). *Survey sampling theory and methods* (1st ed.). Marcel Dekker, N.Y.: CRC Press; (2nd ed.). Florida, N.Y.: USA

Chaudhuri, A. (2014). *Modern survey sampling.* Florida, USA: CRC Press.

Rubin, D. B. (1986). Basic ideas of multiple imputation for nonresponse. *Survey Methodology, 12,* 37–47.

Rubin, D. B. (1988). An overview of multiple imputation. In *Proceedings of the survey research method section* (pp. 79–84). American Statistical Association.

Chapter 6
Randomized Response and Indirect Survey Techniques

6.1 Warner (1965): His Innovation

Warner (1965) introduced the pioneering technique of Randomized Response (RR). He addressed only the case of qualitative sensitive characteristics in a dichotomous situation. He supposed that in a human population every person either bears a sensitive characteristic A or its complement A^C. Though he did not clarify explicitly, the characteristic A^C he conceived is apparently innocuous. For example, A stands for alcoholism or ill-treatment of the spouse or testing HIV positive or being an AIDS patient, etc.

Essentially, his randomized data-gathering device is the following. A sampled person labelled i, how so ever selected, is approached by an investigator with a box of a number of identical-looking cards marked either A or A^C in proportions $p : (1 - p)$ with $0 < p < 1$ but $p \neq \frac{1}{2}$.

Then, unseen by the investigator, the sampled person is to randomly draw a card from the box and respond 'Yes' if the mark A or A^C on the card 'matches' his/her feature A or A^C and 'No' if it 'does not match' and put the card back in the box. The purpose of the survey is to estimate suitably the proportion θ of the people in the population bearing A. According to the nature of this device, the respondent's true feature A or A^C is not revealed and so it is expected that the sampled person may have no objection to answering 'Yes' or 'No' with their data-gathering procedure.

Yet it is possible to utilize these responses independently gathered from a number of sampled persons quite easily as follows to unbiasedly estimate θ deriving an expression for the variance of the estimator along with an unbiased estimator thereof.

Also, it is possible to reasonably measure if the respondent's identity has been revealed to any extent and also to revise this 'Warner-device' so that a respondent may be permitted an 'option' to give out a Direct Response if the feature A or its complement A^C does not seem stigmatizing to him/her even though the investigator feels so. Such a story we intend to narrate here in what follows in a comprehensive way.

© The Author(s), under exclusive license to Springer Nature Singapore Pte Ltd. 2022   133
A. Chaudhuri and S. Pal, *A Comprehensive Textbook on Sample Surveys*, Indian Statistical Institute Series, https://doi.org/10.1007/978-981-19-1418-8_6

To proceed analytically, let

$$y_i = 1 \text{ if } i\text{th person bears } A$$
$$= 0 \text{ if } i\text{th person bears } A^C.$$

Let $Y = \sum_1^N y_i$ and $\theta = \frac{Y}{N}$ which we wish to estimate.

Let $I_i = 1$ if for Warner's device i-th person's true feature 'matches' the card type
$= 0$ if it 'does not match'.

Writing E_R, V_R as operators for expectation and variance with respect to any RR technique including Warner's, we get

$$E_R(I_i) = py_i + (1 - p)(1 - y_i) = (1 - p) + (2p - 1)y_i.$$

So, $r_i = \dfrac{I_i - (1 - p)}{(2p - 1)}$ has $E_R(r_i) = y_i$

and $V_R(r_i) = V_i$, say, $= \dfrac{p(1 - p)}{(2p - 1)^2}$ because

$I_i^2 = I_i$ and $y_i^2 = y_i \; \forall \, i \in U.$

Let, from the population, $U = (1, \ldots, i, \ldots, N)$, a sample s be chosen according to a design p with probability $p(s)$. Let E_p, V_p, C_p denote operators according to a design p, for expectation, variance, covariance. Further, let $E = E_p E_R = E_R E_p$ and $V = E_p V_R + V_p E_R = E_R V_p + V_R E_p$, the over-all expectation, variance operators with respect to sample selection and RR-based data gathering.

Let $t = t(s, \underline{Y})$, denote an estimator for $Y = \sum_1^N y_i$ based on sample s chosen according to a design p and $\underline{Y} = (y_1, \ldots, y_N)$.

Let $E_p(t) = Y$, $\underline{R} = (r_1, \ldots, r_N)$ and $e = e(s, \underline{R})$ be an analogous estimator for $R = \sum_1^N r_i$ i.e., it is t above but $\underline{Y}$ replaced by $\underline{R}$, based on s such that $E_p(e) = R$. As a matter of fact, $e = t(s, \underline{Y})|_{\underline{Y}=\underline{R}}$.

For illustration, we shall consider for p and t the sampling designs and schemes and the estimators we have illustrated in our Chaps. 1–3.

Thus, let t be taken as the Horvitz–Thompson (HT, 1952) estimator, namely

$$t_{HT} = \sum_{i \in s} \frac{y_i}{\pi_i}, \; \pi_i > 0 \, \forall \, i \in U. \text{ Then}$$

$$V_p(t_{HT}) = \sum_{i<j}^N \sum^N (\pi_i \pi_j - \pi_{ij}) \left(\frac{y_i}{\pi_i} - \frac{y_j}{\pi_j} \right)^2 + \sum_1^N \frac{y_i^2}{\pi_i} \alpha_i$$

with $\alpha_i = 1 + \dfrac{1}{\pi_i} \displaystyle\sum_{\substack{j=1 \\ j \neq i}}^{N} \pi_{ij} - \sum_{1}^{N} \pi_i$

and $\widehat{V_p}(t_{HT}) = v_p = \displaystyle\sum\sum_{i<j\in s} \dfrac{(\pi_i \pi_j - \pi_{ij})}{\pi_{ij}} \left(\dfrac{y_i}{\pi_i} - \dfrac{y_j}{\pi_j}\right)^2 + \sum_{i\in s} \dfrac{y_i^2}{\pi_i^2}\alpha_i,$ taking π_{ij}

$> 0 \,\forall i \neq j \in U$

is an unbiased estimator of $V_p(t_{HT})$.

Now, if RR data are gathered for such a sample by Warner (1965) device, then we may consider employing for 'Y' the estimator

$$e_{HT} = \sum_{i\in s} \frac{r_i}{\pi_i}.$$

Then, $E(e_{HT}) = E_p(t) = Y$, ie, e_{HT} is unbiased for Y.

Then, $V(e_{HT}) = \displaystyle\sum_{1}^{N} \frac{V_i}{\pi_i} + \sum_{i<j}^{N}\sum^{N}(\pi_i \pi_j - \pi_{ij})\left(\frac{y_i}{\pi_i} - \frac{y_j}{\pi_j}\right)^2 + \sum \frac{y_i^2}{\pi_i}\alpha_i$

$= \dfrac{p(1-p)}{(2p-1)^2} \displaystyle\sum_{1}^{N} \frac{1}{\pi_i} + \sum_{i<j}^{N}\sum^{N}(\pi_i\pi_j - \pi_{ij})\left(\frac{y_i}{\pi_i} - \frac{y_j}{\pi_j}\right)^2 + \sum_{1}^{N} \frac{y_i^2}{\pi_i}\alpha_i$

$= \dfrac{p(1-p)}{(2p-1)^2} \displaystyle\sum_{1}^{N} \frac{1}{\pi_i} + V_p(t).$

In order to device an unbiased estimator for this $V(e_{HT})$, let us start with

$$a = \sum\sum_{i<j\in s} \frac{\pi_i \pi_j - \pi_{ij}}{\pi_{ij}}\left(\frac{r_i}{\pi_i} - \frac{r_j}{\pi_j}\right)^2 + \sum_{i\in s}\frac{r_i^2}{\pi_i^2}\alpha_i.$$

Then

$$E_R\left(\frac{r_i}{\pi_i} - \frac{r_j}{\pi_j}\right)^2 = E_R\left[\frac{r_i - y_i}{\pi_i} - \frac{r_j - y_j}{\pi_j} + \left(\frac{y_i}{\pi_i} - \frac{y_j}{\pi_j}\right)\right]^2$$

$$= \left(\frac{y_i}{\pi_i} - \frac{y_j}{\pi_j}\right)^2 + \left(\frac{V_i}{\pi_i^2} + \frac{V_j}{\pi_j^2}\right). \text{ Also,}$$

$$E_R\left(\frac{r_i^2}{\pi_i^2}\alpha_i\right) = \frac{y_i^2}{\pi_i^2}\alpha_i + \frac{V_i^2}{\pi_i^2}\alpha_i.$$

$$\text{So, } E_R(a) = \sum\sum_{i<j\in s}\left(\frac{\pi_i\pi_j - \pi_{ij}}{\pi_{ij}}\right)\left(\frac{y_i}{\pi_i} - \frac{y_j}{\pi_j}\right)^2 + \sum_{i\in s}\frac{y_i^2}{\pi_i^2}\alpha_i$$

$$+ \sum\sum_{i<j\in s}\left(\frac{\pi_i\pi_j - \pi_{ij}}{\pi_{ij}}\right)\left(\frac{V_i}{\pi_i^2} + \frac{V_j}{\pi_j^2}\right) + \sum_{i\in s}\frac{V_i}{\pi_i^2}\alpha_i.$$

$$\text{So, } \qquad E_R(a) = \sum\sum_{i<j\in s}\left(\frac{\pi_i\pi_j - \pi_{ij}}{\pi_{ij}}\right)\left(\frac{y_i}{\pi_i} - \frac{y_j}{\pi_j}\right)^2 + \sum_{i\in s}\frac{y_i^2}{\pi_i^2}\alpha_i$$

$$+ \sum\sum_{i<j\in s}\left(\frac{\pi_i\pi_j - \pi_{ij}}{\pi_{ij}}\right)\left(\frac{V_i}{\pi_i^2} + \frac{V_j}{\pi_j^2}\right) + \sum_{i\in s}\frac{V_i}{\pi_i^2}\alpha_i$$

$$\text{So, } \qquad E(a) = \sum_{i<j}^{N}\sum^{N}(\pi_i\pi_j - \pi_{ij})\left(\frac{y_i}{\pi_i} - \frac{y_j}{\pi_j}\right)^2 + \sum_1^N\frac{y_i^2}{\pi_i}\alpha_i$$

$$+ \sum_{i<j}^{N}\sum^{N}(\pi_i\pi_j - \pi_{ij})\left(\frac{V_i}{\pi_i^2} + \frac{V_j}{\pi_j^2}\right) + \sum_1^N\frac{V_i}{\pi_i}\alpha_i$$

$$\text{i.e., } \qquad E(a) = \sum_1^N\frac{V_i}{\pi_i} + V_p(t) - \sum V_i$$

$$\text{So, } \qquad v(e) = a + \sum_{i\in s}\frac{V_i}{\pi_i} \text{ has } E[v(e)] = V(e).$$

$$\text{Thus, } \quad v(e) = \frac{p(1-p)}{(2p-1)^2}\sum_{i\in s}\frac{1}{\pi_i} + \sum\sum_{i<j\in s}\frac{(\pi_i\pi_j - \pi_{ij})}{\pi_{ij}}\left(\frac{r_i}{\pi_i} - \frac{r_j}{\pi_j}\right)^2 + \sum_{i\in s}\frac{r_i^2}{\pi_i^2}\alpha_i$$

is an unbiased estimator for $V(e)$.

Warner (1965), however, illustrated only the use of the sample mean of all the units selected by SRSWR to unbiasedly estimate $\theta = \bar{Y}$, the population mean or the proportion bearing A.

Thus, corresponding to the estimator $\bar{y} = \frac{1}{n}\sum_{K=1}^{n} y_K$ with y_K as the y_i-value for the unit chosen on the K-th $(K = 1, \ldots, n)$ draw in an SRSWR in n draws from the population, for the unbiased estimator in a direct survey (DR) for the population mean $\bar{Y} = \frac{1}{N}\sum_1^N y_i = \theta$, say, his estimator based on the RR survey for the SRSWR taken in n draws is $\bar{r} = \frac{1}{n}\sum_1^n r_k$. Here r_k be the value of r_i for the unit chosen on the K-th draw as the value $r_K = \frac{I_R - (1-p)}{(2p-1)}$ for his RR device with $I_K = 1$, if the card type A or A^C 'matches' the feature of the person chosen on the Kth draw exercising Warner's RR device

$= 0$, if it 'does not match'.

$$\text{Then, } E_R(I_K) = py_k + (1-p)(1-y_K),$$
$$E_R(r_k) = py_k \text{ and}$$
$$V_R(r_k) = \frac{p(1-p)}{(2p-1)^2} = V_K, \text{ say}$$

$k = 1, \ldots, n.$

Then the RR-based estimator for θ, viz.

$$\bar{r} \text{ has } E_R(\bar{r}) = \bar{y}, \; V_R(\bar{r}) = \frac{1}{n^2} \sum_{K=1}^{n} V_K$$

$$\text{or } V_k(\bar{r}) = \frac{p(1-p)}{n(2p-1)^2} \; \forall k = 1, \ldots, n.$$

$$\text{So, } E(\bar{r}) = E_p(\bar{y}) = \bar{Y} = \theta \text{ and}$$

$$V(\bar{r}) = V_p(\bar{y}) + E_p V_K(\bar{y})$$
$$= \frac{\sigma^2}{n} + \frac{p(1-p)}{n(2p-1)^2}, \text{ where}$$

$$\sigma^2 = \frac{1}{N} \sum_{i=1}^{N} (y_i - \bar{Y})^2 = \frac{1}{N} \left[\sum_{1}^{N} y_i^2 - N(\bar{Y})^2 \right]$$

$$= \frac{1}{N} \left[\sum_{1}^{N} y_i - N\theta^2 \right] = \theta(1-\theta). \text{ So,}$$

$$V(\bar{r}) = \frac{1}{n} \left[\theta(1-\theta) + \frac{p(1-p)}{(2p-1)^2} \right]$$

On the other hand

$$V(\bar{y}) = \frac{\sigma^2}{n} = \frac{\theta(1-\theta)}{n}.$$

An unbiased estimator for $V(\bar{r})$ is

$$v = \frac{1}{(n-1)} \left[\bar{r}(1-\bar{r}) + \frac{p(1-p)}{(2p-1)^2} \right]$$

$$\text{because } E_R(v) = \frac{1}{(n-1)} \left[\bar{y} - \frac{p(1-p)}{n(2p-1)^2} - \bar{y}^2 + \frac{p(1-p)}{(2p-1)^2} \right]$$

$$= \frac{1}{(n-1)} \left[\bar{y}(1-\bar{y}) + \frac{p(1-p)(n-1)}{n(2p-1)^2} \right]$$

$$\text{and } E(v) = E_p E_R(v) + V_p(\bar{y}) = \frac{\theta(1-\theta)}{n} + \frac{p(1-p)}{(2p-1)^2} \frac{1}{n} = V(\bar{r})$$

A question now is whether the privacy of the respondent is totaly protected or may be revealed and if so, to what extent.

Let us write L_i to denote an unknown likelihood that a person labelled i bears the feature A.

For Warner (1965) RRT we have

$$P_r(I_i = 1 | y_i = 1) = p, \ P_r(I_i = 1 | y_i = 0) = (1-p)$$
$$P_r(I_i = 0 | y_i = 1) = (1-p), \ P_r(I_i = 0 | y_i = 0) = p$$

Then, by Bayes' theorem, we may write the posterior probabilities that a person i may bear A or bear A^C, once the RR is given as 'yes', i.e., $I_i = 1$ and 'No', i.e., $I_i = 0$ respectively as

$$L_i(1) = \frac{L_i P_r(I_i = 1 | y_i = 1)}{L_i P_r(I_i = 1 | y_i = 1) + (1 - L_i) P_r(I_i = 1 | y_i = 0)}$$

$$= \frac{p L_i}{(1-p) + (2p-1) L_i}$$

The ith person's 'privacy' is protected, the closer the value $L_i(1)$ to $L_i \ \forall i$. Similarly,

$$L_i(0) = \frac{L_i P_r(I_i = 0 | y_i = 1)}{L_i P_r(I_i = 0 | y_i = 1) + (1 - L_i) P_r(I_i = 0 | y_i = 0)}$$

$$= \frac{L_i(1-p)}{L_i(1-p) + (1 - L_i)p} = \frac{L_i(1-p)}{p + (1-2p) L_i}$$

and the closer $L_i(0)$ to L_i the better the privacy is protected.

From the above it is easy to see that $L_i(1) \to L_i$ if $p \to \frac{1}{2}$ and $L_i(0) \to L_i$ if $p \to \frac{1}{2}$.

But as $p \to \frac{1}{2}$, $\frac{p(1-p)}{(2p-1)^2} \to \infty$.

Thus if 'privacy' is better protected then efficiency of the Warner' RRT declines because

$$V(e_{HT}) = \frac{p(1-p)}{(2p-1)^2} \frac{1}{N} \sum_1^N \frac{1}{\pi_i} + V_p(t_{HT})$$

for a general sampling scheme combined with Horvitz-Thompson estimator and

$$V(\bar{r}) = \frac{p(1-p)}{n(2p-1)^2} + \frac{\theta(1-\theta)}{n}$$

if for SRSWR the sample mean of RR's is used to estimate θ.

Thus 'efficiency' level and 'privacy' protection move in opposite directions. But both efficiency and privacy protection are crucial for a successful RRT. So, 'p' should be cleverly chosen to balance the opposite desiderata.

Next we consider Optional RRT's. Suppose even though the researcher in using RR presumes a particular attribute A as stigmatizing a respondent may feel to the contrary and may eagerly give out truthfully that he/she bears this A. Then the investigator may permit the respondent an option either to respond directly if he/she chooses to do so or give him/her the option to adopt the RR device presented. Let us see what may happen if the Warner's RR device may be the option versus the DR.

Suppose in a probability sample s howsoever chosen, a part s_1 consists of those who 'respond directly' about bearing A and the complementary part $s_2 = s - s_1$ respond following the Warner's RR device.

Then, instead of DR-based estimator $t = t(s, \underline{Y})$ for Y and the compulsorily RR-based/eRR-based estimator

$$e = t(s, \underline{Y})|_{\underline{Y}=\underline{R}} = e(s, \underline{R})$$

one should employ the 'Optionally RR-based (ORR)' estimator

$$e^* = t(s_1, \underline{Y}) + e(s_2, \underline{R}) \text{ for } Y.$$

More specifically let

$$t = t(s, \underline{Y}) = \sum_{i \in s} \frac{y_i}{\pi_i},$$

$$e = e(s, \underline{R}) = \sum_{i \in s} \frac{r_i}{\pi_i}, \; y_i = E_R(r_i), \; V_R(r_i) = V_i$$

$$e^* = t(s_1, \underline{Y}) + e(s_2, \underline{R}) = \sum_{i \in s_1} \frac{y_i}{\pi_i} + \sum_{i \in s_2} \frac{r_i}{\pi_i}.$$

Let E_{OR} denote operator for expectation when 'option' is allowed for DR or RR.

$$\text{Then, } E_{OR}(e) = \sum_{i \in s_1} \frac{y_i}{\pi_i} + \sum_{i \in s_2} \frac{r_i}{\pi_i} = e^*$$

$$E_R(e^*) = t = E_R(e), \; V_R(e^*) = E_R(e^* - t)^2$$

$$\text{Now, } E_R(e-e^*)^2 = E_R\left[(e-t) - (e^*-t)\right]^2$$
$$= V_R(e) + V_R(e^*) - 2E_R(e^*-t)E_{OR}(e-t)$$
$$= V_R(e) + V_R(e^*) - 2E_R(e^*-t)^2$$
$$= V_R(e) - V_R(e^*)$$

So,
$$V_R(e^*) = V_R(e) - E_R(e-e^*)^2$$
$$\leq V_R(e) \text{ unless } e \text{ coincides with } e^*.$$

So,
$$V(e^*) = E_p V_R(e^*) + V_p E_R(e^*)$$
$$= E_p V_R(e) - E_p E_R(e-e^*)^2 + V_p E_R(e^*)$$
$$= E_p V_R(e) - E_p E_R(e-e^*)^2 + V_p E_R(e)$$
$$= V(e) - E_p E_R(e-e^*)^2.$$

So, an unbiased estimator for $V(e^*)$ is

$$v(e^*) = v(e) - (e-e^*)^2$$
$$= \frac{p(1-p)}{(2p-1)^2}\sum_{i\in s}\frac{1}{\pi_i} + \sum_{i<j\in s}\sum\left(\frac{\pi_i\pi_j - \pi_{ij}}{\pi_{ij}}\right)\left(\frac{r_i}{\pi_i} - \frac{r_j}{\pi_j}\right)^2$$
$$- \left[\sum_{i\in s_1}\left(\frac{r_i - y_i}{\pi_i}\right)\right]^2$$

Since $E_R(e-e^*)^2 = \frac{p(1-p)}{(2p-1)^2}\sum_{i\in s_1}\frac{1}{\pi_i^2}$ a second unbiased estimator of $V(e)$ is

$$v'(e) = \frac{p(1-p)}{(2p-1)^2}\sum_{i\in s}\frac{1}{\pi_i} + \sum_{i<j\in s}\sum\left(\frac{\pi_i\pi_j - \pi_{ij}}{\pi_{ij}}\right)\left(\frac{r_i}{\pi_i} - \frac{r_j}{\pi_j}\right)^2 - \frac{p(1-p)}{(2p-1)^2}\sum_{i\in s_1}\frac{1}{\pi_i^2}.$$

An interested reader may see Chaudhuri and Mukerjee (1988, 1985) for optional randomized response literature.

6.2   More RR Devices to Estimate Population Proportion

(i) Unrelated Question Device: $\underline{A}$, $\underline{A^C}$ both sensitive
Simmons, vide (Horvitz et al., 1967; Greenberg et al., 1969), essentially gives the following URL device enabling us to take care of both A and its complement A^C allowed to be stigmatizing. Every probability-sampled person labelled i, no matter how, is approached with 2 boxes of cards labelled A and B in proportions $p_1 : (1 - p_1)$ in one and $p_2 : (1 - p_2)$ in the other ($0 < p_1 < 1$ and $0 < p_2 <$

1, $p_1 \neq p_2$). Here A is a feature suspected to be sensitive but B is an unrelated innocuous feature like for example, A denoting 'torturous to spouse's and B denoting 'preference for music to sport'. Every chosen person however selected is requested independently of each other to take a card from the 1-st box to respond

$$I_i = 1 \text{ if the card type } A \text{ or } B \text{ drawn}$$
$$\text{'matches' the person's corresponding feature}$$
$$= 0 \text{ if 'it does not match'}$$

and the card is returned to the box without divulging to the investigator the card type and the actual feature.
A similar exercise independently with the second box yields the RR as

$$J_i = 1 \text{ if the card type } A \text{ or } B \text{ 'matches' the person's } A \text{ or } B \text{ feature}$$
$$= 0 \text{ if 'it does not match'.}$$

$$y_i = 1 \text{ if } i \text{ bears } A$$
$$= 0 \text{ if } i \text{ bears } A^C.$$

$$x_i = 1 \text{ if } i \text{ bears } B$$
$$= 0 \text{ if } i \text{ bears } B^C, \text{ the complement of } B.$$

Then, similarly to Warner's RR device follows

$$E_R(I_i) = p_1 y_i + (1 - p_1)x_i$$
$$E_R(J_i) = p_2 y_i + (1 - p_2)x_i \text{ giving us}$$
$$r_i = \frac{(1 - p_2)I_i - (1 - p_1)J_i}{(p_1 - p_2)} \text{ with } E_R(r_i) = y_i.$$

Also,

$$V_R(r_i) = E_R(r_i^2) - [E_R(r_i)]^2$$
$$= E_R(r_i^2) - y_i^2 = E_R(r_i^2) - y_i$$
$$= E_R(r_i^2) - E_R(r_i) = E_R[r_i(r_i - 1)]$$
$$= V_i(\text{ say}).$$

So, $r_i(r_i - 1) = v_i$ may be taken as an unbiased estimator for V_i. Since, apparently $r_i(r_i - 1)$ may often turn out negative, we may find out as follows an alternative unbiased estimator for the above variance V_i too.

Now, $V_i = V_R(r_i) = \dfrac{(1 - p_2)^2 V_R(I_i) + (1 - p_1)^2 V_R(J_i)}{(p_1 - p_2)^2};$

$$V_R(I_i) = E_R(I_i)(1 - E_R(I_i)) \text{ since } I_i^2 = I_i$$
$$= p_1(1 - p_1)(y_i - x_i)^2 \text{ since } y_i^2 = y_i,\, x_i^2 = x_i.$$

Similarly, $V_R(J_i) = p_2(1 - p_2)(y_i - x_i)^2$ and

$$V_i = V_R(r_i) = \frac{(1 - p_1)(1 - p_2)(p_1 + p_2 - 2 p_1 p_2)}{(p_1 - p_2)^2}(y_i - x_i)^2$$

Now, for $Y = \displaystyle\sum_1^N y_i = N\theta$ an unbiased estimator on $Y_i,\, i \in s$ may be taken as

$$e_H = \sum_{i \in s} \frac{r_i}{\pi_i} \text{ analogously to } t = \sum_{i \in s} \frac{y_i}{\pi_i}$$

if DR's y_i's were available for i in s.

Then, $V(e_H) = E_p \left(\displaystyle\sum_{i \in s} \frac{V_i}{\pi_i^2} \right) + \displaystyle\sum_{i<j}^N \sum^N (\pi_i \pi_j - \pi_{ij}) \left(\frac{y_i}{\pi_i} - \frac{y_j}{\pi_j} \right)^2 + \displaystyle\sum_1^N \frac{y_i^2}{\pi_i} \beta_i$

with $\beta_i = 1 + \dfrac{1}{\pi_i} \displaystyle\sum_{\substack{j=1 \\ j \neq i}}^N \pi_{ij} - \sum_1^N \pi_i$

or $V(e_H) = \displaystyle\sum_1^N \frac{V_i}{\pi_i} + \displaystyle\sum_{i<j}^N \sum^N (\pi_i \pi_j - \pi_{ij}) \left(\frac{y_i}{\pi_i} - \frac{y_j}{\pi_j} \right)^2 + \displaystyle\sum_1^N \frac{y_i^2}{\pi_i} \beta_i.$

Then, an unbiased estimator for this $V(e_H)$ is

$$v(e_H) = \sum_{i \in s} \frac{w_i}{\pi_i} + \sum_{i<j \in s} \sum \left(\frac{\pi_i \pi_j - \pi_{ij}}{\pi_{ij}} \right) \left(\frac{r_i}{\pi_i} - \frac{r_j}{\pi_j} \right)^2 + \sum_{i \in s} \frac{r_i^2}{\pi_i^2} \beta_i, \text{ taking } \pi_{ij} > 0 \,\forall\, i, j$$

Here w_i is taken as an unbiased (approximately) estimator for $V_i = V_R(r_i)$, writing

$$\frac{(1 - p_1)(1 - p_2)(p_1 + p_2 - 2 p_1 p_2)}{(p_1 - p_2)^2} \text{Est}(y_i - x_i)^2$$

writing $\text{Est}(y_i - x_i)^2$ as

$$(r_i - t_i)^2 - r_i(r_i - 1) - t_i(t_i - 1)$$

and $t_i = \dfrac{p_1 J_i - p_2 I_i}{p_1 - p_2}$ for which

$$E_R(t_i) = x_i$$

$$V_R(t_i) = E_R[t_i(t_i - 1)] \ cf \ V_R(r_i) = E_R[r_i(r_i - 1)]$$

Also, $E_R(r_i - t_i)^2 = E_R[(r_i - y_i) - (t_i - x_i) + (y_i - x_i)]^2$

So, $\text{Est}(y_i - x_i)^2 = (r_i - t_i)^2 - r_i(r_i - 1) - t_i(t_i - 1)$

neglecting $E_R(r_i - y_i)(t_i - x_i)$

This is easy to check analogously to the case of RR's gathered by Warner's RR device.

For this URL device also an optional RR version ORR and how to examine protection of privacy may be followed through with little difficulty as studied in case of Warner's device.

(ii) Kuk's (1990) RR Device

Suppose a person labelled i has been chosen by some probability sampling procedure.

The interviewer is to approach him/her with 2 boxes. In the 1st box there are 'red' and 'non-red' cards in proportions $\theta_1 : (1 - \theta_1)$, $(0 < \theta_1 < 1)$ and in the other box they are in proportions $\theta_2 : (1 - \theta_2)$, $(0 < \theta_2 < 1)$ but $(\theta_1 \neq \theta_2)$. The sampled person bearing A is to draw from the 1-st box cards by SRSWR a number of times K and report the number f_i of times a 'red' card has been observed; if the person bears A^C, the same exercise is to be performed using the 2nd box. The person is not to divulge his/her feature A or A^C and not to disclose also the box he/she used.

Then will follow the following.

$$E_R(f_i) = K[y_i\theta_1 + (1 - y_i)\theta_2] \text{ and}$$
$$V_R(f_i) = K[y_i\theta_1(1 - \theta_1) + (1 - y_i)\theta_2(1 - \theta_2)]$$

with of course, $y_i = 1$ if i bears A

$$= 0 \text{ if } i \text{ bears } A^C.$$

Consequently,

$$r_i(K) = \frac{f_i/K - \theta_2}{\theta_1 - \theta_2} \text{ has } E_R r_i(K) = y_i$$

and $V_R r_i(K) = \dfrac{V_R(f_i)}{K^2(\theta_1 - \theta_2)^2} = V_i(K)$

$$= a(K) + b(K)y_i, \text{ writing}$$

$$a(K) = \frac{\theta_2(1 - \theta_2)}{K^2(\theta_1 - \theta_2)^2} \text{ and } b(K) = \frac{(1 - \theta_1 - \theta_2)}{K^2(\theta_1 - \theta_2)^2}$$

So, $v_i(K) = a(K) + b(K)r_i(K)$

is an unbiased estimator for $V_R r_i(K)$.

Hence follows similarly the follow-up similarly what we dealt with concerning Warner's and Simmon's RR device.

(iii) Forced Response Device

A person labelled $i = (1, \ldots, N)$ selected by any probability sampling procedure is approached with a box of a large number of identical cards distinguished by marks 'yes', 'no' and 'genuine' in proportions $p_1, p_2, 1 - p_1 - p_2 (0 < p_1, p_2 < 1, p_1 \neq p_2, 1 - p_1 - p_2 > 0)$. On request he/she is to draw one card randomly to respond

$I_i = 1$ if card is marked 'yes' or 'genuine' and his/her feature is A

$\quad = 0$ if card is marked 'no' or 'genuine' but his/her feature is not A i.e., A^C.

$$\text{As usual } y_i = 1 \text{ if } i \text{ bears } A$$
$$= 0 \text{ if } i \text{ bears } A^C.$$

Then, $\text{Prob}(I_i = 1 | y_i = 1) = p_1 + (1 - p_1 - p_2) = 1 - p_2$

$\text{Prob}(I_i = 0 | y_i = 1) = p_2$, $\text{Prob}(I_i = 1 | y_i = 0) = p_1$

$\text{Prob}(I_i = 0 | y_i = 0) = p_2 + (1 - p_1 - p_2) = 1 - p_1$

$$\text{So, } E_R(I_i) = \text{Prob}(I_i = 1) = p_1 + y_i(1 - p_1 - p_2)$$
$$\text{Then, } r_i = \frac{I_i - p_1}{1 - p_1 - p_2} \text{ has } E_R(I_i) = y_i \text{ and}$$

$$V_R(r_i) = V_i = \frac{V_R(I_i)}{(1 - p_1 - p_2)^2} = \frac{E_R(I_i)(1 - E_R(I_i))}{(1 - p_1 - p_2)^2}$$
$$= \frac{p_1(1 - p_1) + y_i(1 - p_1 - p_2)(p_2 - p_1)}{(1 - p_1 - p_2)^2}$$

since $y_i^2 = y_i \, \forall \, i$.

Hence $\qquad\qquad V_i = p_1(1 - p_1)/(1 - p_1 - p_2)^2$ if $y_i = 0$

$$= \frac{p_2(1 - p_2)}{(1 - p_1 - p_2)^2} \text{ if } y_i = 1$$

But an unbiased estimator for V_i is

$$v_i = \frac{p_1(1 - p_1) + r_i(1 - p_1 - p_2)(p_2 - p_1)}{(1 - p_1 - p_2)^2}$$

So, what was discussed about Warner's and Simmon's devices follow in respect of this 'Forced Response' device also.

Chaudhuri and Roy (1997) studied Model assisted survey sampling strategies with randomized responses. Chaudhuri and Pal (2015) examined improvement by Empirical Bayes procedure, exploiting available data on a correlated variable while using a randomized response technique to gather data on the main variable of interest.

6.3   When the Stigmatizing Feature is Quantitative

Suppose we are interested to estimate the total number of people in a community who are 'AIDS-patients' or to estimate the average number of days spent in detention by the delinquent in a community under a specific category of charges. Thus y is a real-valued variable of interest with values $y_i (i = 1, \ldots, N)$ and we intend to estimate
$$Y = \sum_1^N y_i \text{ or } \bar{Y} = \frac{Y}{N}.$$

Then we may employ one of the two devices A and B described below:

(i) <u>Device A</u>

Let s be a sample chosen according to a design p from a population $U = (1, \ldots, N)$ with probability $p(s)$ and $\pi_i = \sum_{s \ni i} p(s)$ and $\pi_{ij} = \sum_{s \ni i,j} p(s) > 0 \forall i, j (i \neq j)$ in U.

Let a sampled unit labelled i be approached with 2 boxes one of which contains a large number of cards numbered $a_1, \ldots, a_M$ with different frequencies and mean $\mu_a = \frac{1}{M} \sum_1^M a_i \neq 0$, variance $\sigma_a^2 = \frac{1}{M} \sum_1^M (a_i - \mu_a)^2 > 0$ and a second box with similar cards numbered $b_1, \ldots, b_T$ with mean $\mu_b = \frac{1}{T} \sum_{K=1}^T b_K$, variance $\sigma_b^2 = \frac{1}{T} \sum_1^T (b_K - \mu_b)^2$

Each sampled person i is requested to randomly take one card independently from each box and to respond
$$z_i = a_j y_i + b_k \text{ without disclosing } y_i, a_j, b_k$$

Then, $E_R(z_i) = y_i \mu_a + \mu_b$ and
$$V_R(z_i) = y_i^2 \sigma_a^2 + \sigma_b^2. \text{ Then,}$$
$$r_i = \frac{z_i - \mu_b}{\mu_a} \text{ has } E_R(r_i) = y_i,$$
$$V_R(r_i) = y_i^2 \left(\frac{\sigma_a^2}{\mu_a^2}\right) + \left(\frac{\sigma_b^2}{\mu_a^2}\right) = V_i, \text{ say.}$$

Then, $v_i = \dfrac{r_i^2 \left(\frac{\sigma_a^2}{\mu_a^2}\right) + \left(\frac{\sigma_b^2}{\mu_a^2}\right)}{1 + \frac{\sigma_a^2}{\mu_a^2}} \text{ has } E_R(v_i) = V_i.$

(ii) Device B
A sampled person i as in Device A is offered a box of cards, a proportion $C(0 < C < 1)$ of which are marked with 'Correct' and the remaining proportion $(1 - C)$ of them are marked $x_1, \ldots, x_j, \ldots, x_T$ in proportions $q_1, \ldots, q_j, \ldots, q_T$ such that $(0 < q_j < 1 \forall j$ and $\sum_1^T q_j = 1 - C)$. The person on request is to draw randomly just one card and is to truthfully report his/her true value y_i if a 'correct' marked card is drawn or the value x_j if an x_j-marked card is drawn without reporting the type of card is marked and is to put it back into the box. The x_j's are cleverly chosen so as to cover the possible value y_i and the respondent is asked to report y_i as a number similar to x_j's, say, correct to the same number of decimal places.

Then, $E_R(z_i) = Cy_i + \sum_1^T q_j x_j$ and

$$r_i = \frac{z_i - \sum_1^T q_j x_j}{C} \text{ has } E_R(r_i) = y_i,$$

$$V_R(r_i) = \frac{1}{C^2}\left[C(1-C)y_i^2 + \sum_1^T q_j x_j^2 - \left(\sum_1^T q_j x_j\right)^2\right]$$

$$= \alpha y_i^2 + \beta y_i + \psi \text{ collecting the coefficients}$$

$$= V_i, \text{ say,}$$

and $v_i = \dfrac{\alpha r_i^2 + \beta r_i + \psi}{1 + \alpha}$, (of course, $1 + \alpha \neq 0$)

Then, estimation, using RR data by both these devices, of $Y, \bar{Y}$, along with unbiased variance estimation follows easily as with RR devices by Warner, Simmons and other.

(iii) Optional RR on quantitative feature
Suppose a probability sample-selected person labelled $i(= 1, \ldots, N)$ is asked at pleasure to choose a probability C_i using a random number table and without disclosing this C_i and that he/she is using it, give out with this probability his/her true value y_i or with the complementary probability $(1 - C_i)$ apply the Device A above supplemented by the addition of a third box of cards marked $b'_1, \ldots, b'_L$ with $\mu'_b = \frac{1}{L}\sum_1^L b'_i \neq \mu_b$ with the further stipulation that $\mu_a = 1$ and let the optional RR from i be $z_i = y_i$ with probability C_i or

$$I_i = a_j y_i + b_k (j = 1, \ldots, M; k = 1, \ldots, T)$$

with probability $(1 - C_i)$; further, let independently a second ORR from i be

$$z_i' = y_i \text{ with probability } C_i$$

or, $I_i' = a_l y_i + b_u' (l = 1, \ldots, M; u = 1, \ldots, L)$ with probability $(1 - C_i)$

Then,
$$E_R(z_i) = C_i y_i + (1 - C_i)(y_i \mu_a + \mu_b)$$
$$= C_i y_i + (1 - C_i)(y_i + \mu_b)$$

$$\text{and } E_R(z_i') = C_i y_i + (1 - C_i)(y_i + \mu_b')$$
$$\text{Hence, } E_R(\mu_b' z_i - \mu_b z_i') = (\mu_b' - \mu_b) y_i$$

$$\text{and } r_{1i} = \frac{(\mu_b' z_i - \mu_b z_i')}{(\mu_b' - \mu_b)} \text{ has } E_R(r_{1i}) = y_i.$$

Now repeating the above exercise one may check that another independent estimate r_{2i} may be obtained with $E_R(r_{2i}) = y_i$.
Then $r_i = \frac{1}{2}(r_{1i} + r_{2i})$ with $E_R(r_i) = y_i$ may be generated yielding $\frac{1}{4}(r_{1i} - r_{2i})^2$ as an unbiased estimator v_i for $V_R(r_i) = V_i$, say.
Now an unbiased estimator for Y, $\bar{Y}$ along with unbiased variance estimator based on z_i, z_i' independently repeated once again may be easily derived.

(iv) Protection of Privacy in Quantitative RR

Let us consider Device A. The quantitative RR from i-th person is

$$z_i = a_j y_i + b_k; \ j = 1, \ldots, M; k = 1, \ldots, T.$$

Each of these z_i's for a given i arises with

$$\text{Prob}(z_i) = \frac{1}{MT} \text{ for } j = 1, \ldots, M; k = 1, \ldots, T$$

Let $L_i = L(y_i)$ be the unknown prior probability that the ith person's true y-value is $y_i (i = 1, \ldots, N)$. Then, given the observed RR in respect of the i-th person z_i, the posterior probability that his/her true y-value is y_i is

$$L(y_i|z_i) = \frac{L_i P(z_i|y_i)}{P(z_i)}$$

writing $P(z_i)$ as the over-all probability of observing z_i no matter what the underlying true-values y_i's.

$$\text{Also, } P(z_i) = \sum_{y_i} L_i P(z_i|y_i)$$

$$\text{So, } L(y_i|z_i) = \frac{L_i \left(\frac{1}{TM}\right)}{\left(\frac{1}{TM}\right)} = L_i \ \forall i \in U.$$

So, unless $T = 1$ and $M = 1$, the posterior and prior probability for every y-value remains in tact whatever RR as z_i one may observe.

If $T = 1$ and $M = 1$, the true value of $y = y_i$ is revealed from the generated RR as z_i

$$y_i = \frac{z_i - b_1}{a_1}$$

with Probability one.

But if $T = 2$ and $M = 2$, then true value y_i given z_i, is revealed with probability $\frac{1}{4}$ for $\mathrm{Prob}(a_1, b_1) = \mathrm{Prob}(a_1, b_2) = \mathrm{Prob}(a_2, b_1) = \mathrm{Prob}(a_2, b_2) = \frac{1}{4}$.

6.4   Lessons, Exercises and Case Study

Lessons

(i) Christofides's RRT

This is a generalization of Warner's RRT. Here a person i selected by a probability sampling method is offered a box with M cards identically designed but differently marked $1, 2, \ldots, K, \ldots, M$ in proportions $p_1, p_2, \ldots, p_K (0 < p_j < 1, \forall j, \sum_1^K p_j = 1)$. The person is requested to report K if he/she bears A^C and a card marked K is randomly chosen or to report $(M + 1 - K)$ if he/she bears A and draws the card marked K without divulging his/her true feature.

$$\text{Let, } y_i = 1 \text{ if } i \text{ bears } A$$
$$= 0 \text{ if } i \text{ bears } A^C.$$

Then explain how to unbiasedly estimate $\theta = \frac{1}{N} \sum_1^N y_i$ and present an unbiased estimate for the variance of this estimator for a general case and show that if $M = 2$, $p_1 = p$ and $p_2 = 1 - p$, then this reduces to Warner (1965) RRT.

(ii) Mangat (1992) RRT

This is a modification of Simmons's URL model. Here, as usual, for an i in U

$$y_i = 1 \text{ if } i \text{ bears } A$$
$$= 0 \text{ if } i \text{ bears } A^C$$
$$x_i = 1 \text{ if } i \text{ bears } B$$
$$= 0 \text{ if } i \text{ bears } B^C; A, B \text{ are unrelated.}$$

A person i selected by a Probability-sampling method is presented 3 boxes. He/she is to randomly choose from the 1-st box a card from the cards marked 'Direct' or 'RR' in proportions $T : (1 - T), 0 < T < 1$. If a 'Direct' card is drawn his/her response is the value $y_i, i \in U$. If an 'RR' marked card is drawn, he/she is to randomly choose a card from the 2nd box where A and B marked cards are there in proportions $p_1 : (1 - p_1)$. Then, independently of what precedes he/she is to take a card again randomly from the 1st box and give out y_i if a 'Direct'-marked card is drawn or if not, take a card randomly from the 3rd box which contains A and B-marked cards in proportions $p_2 : (1 - p_2), (0 < p_2 < 1$ but $p_2 \neq p_1)$. Denoting the responses as y_i or I_i or J_i with obvious implications, check that

$$E_R(I_i) = Ty_i + (1 - T)[p_1 y_i + (1 - p_1)x_i]$$
$$\text{and } E_R(J_i) = Ty_i + (1 - T)[p_2 y_i + (1 - p_2)x_i]$$

Now find (i) an unbiased estimator for $\theta = \frac{1}{N} \sum^N y_i$ along with (ii) an unbiased estimator for the variance of this estimator.

(iii) Liu, Chow and Mosley's RRT

Suppose every person i in U bears one of T mutually exclusive features in proportions $\pi_1, \ldots, \pi_k, \ldots, \pi_T$ such that $(0 < \pi_k < 1, k = 1, \ldots, T, \sum_1^T \pi_k = 1)$ and a few of them are sensitive and identifiable. Say, feature 1 denotes illegal drivers, 3 denotes Income Tax deceivers, 7 denotes false recipients of Government concessions and so on. Our interest is to estimate certain proportions $\pi_j, j = 1, \ldots, T$.

Liu et al. (1975) gave us the following RRT.

A probability sample-selected person i is then approached with a flask with a transparent long neck and beads of T different colours and numbering $m_1 \neq m_2 \neq \cdots \neq m_T$ are placed inside the flask. The flask has a cork as a stopper so that with it the top of the flask can be closed so that if the beads are placed in the flask they may be shaken at will and when turned upside down and closed with the stopper cork, the beads do not drop out and stand one on another and the neck is marked $1, 2, \ldots, m_k \left(m = \sum_1^T m_k \right)$ and one may note the number at which the bottom-most of a colour denoting a person's feature stands and the mark may be noticed.

Such a number is to be reported by a sampled respondent as his/her RR. Then,

$$p_{11} = \text{Prob[a respondent of category 1 reports 1]} = \frac{m_1}{m}$$

$$p_{23} = \text{Prob[a respondent of category 2 reports 3]}$$

$$= \left(\frac{m - m_2}{m} \right) \left(\frac{m - 1 - m_2}{m - 1} \right) \left(\frac{m_3}{m - 2} \right) \text{ and so on.}$$

Thus we may calculate

$p_{jk} = \text{Prob[a respondent of category } j \text{ reports } k]$ for $j, k = 1, 2, \ldots$

Then,

$$\lambda_j = \text{Prob[an RR is } j\,] = \sum_1^T p_{kj}\pi_k$$

for $j = 1, 2, \ldots, M = \max(m - m_k + 1), 1 < k < T$.

Now let $I_{ij} = 1$ if a sampled person i reports j
$= 0$ if i reports a non-j.

Then, $E_R(I_{ij}) = \lambda_j \, \forall i$ in U.

Let $I'_{ij} = 1$ for an independent 2nd response j
from the same sampled person i
$= 0$ if response is a non-j.

Let $J_{ij} = \dfrac{1}{2}(I_{ij} + I'_{ij})$

Then, $E_R(J_{ij}) = \lambda_j \, \forall i \in U$.

$$V_{ij} = V_R(J_{ij}) = \frac{V_R(I_{ij})}{2}$$

and $v_{ij} = \dfrac{1}{4}\left(I_{ij} - I'_{ij}\right)^2 \, \forall i \in U$

and $E_R(v_{ij}) = V_{ij} \, \forall i \in U$.

Now J_{ij} for $i \in s$ provides an unbiased estimator $\widehat{\lambda}_j$ for λ_j by using say the Horvitz-Thompson estimator $e = \sum\limits_{i \in s} \frac{J_{ij}}{\pi_i}$, writing $\pi_i = \sum\limits_{s \ni i} p(s)$, the inclusion-probability of i according to design p.
Now work out formulae for $V(e)$ and for unbiased estimator for $V(e)$.
By method of moments, using

$$\pi_j = \sum_{k=1}^T p_{pj}\pi_K$$

work out unbiased estimator for π_K along with unbiased variance estimators thereof.

Exercises

(i) Suppose y denotes the 'fine' imposed and realized from a delinquent car driver
   in a community. Describe how you may unbiasedly estimate the total y-value for
   the community by adopting (Rao et al., 1962) sample selection and an optional
   RR technique, along with an unbiased vadiance estimator for it.
   For a clue one may consult pp. 110–111 of Chaudhuri (2011) book.

(ii) Consider Chaudhuri's Randomized Device I for a sensitive quantitative variable
   y, say, denoting the 'fines' imposed and realized against rash motor driving.
   Discuss how a respondent's privacy may be protected at least when the device
   is suitably employed.
   For a clue one may see pp. 164–166 in Chaudhuri and Christofides (2013) book.

(iii) What may be your conclusion if, instead, Chaudhuri's device II is employed.
   For a clue, one may consult pp. 166–167 in the book by Chaudhuri and
   Christofides (2013).

Case Studies

We shall consider here a numerical study of how efficiency levels if three classical
RRT's fare versus their protection of privacy measures. The three RRT's we take up
here are Warner (1965), Simmon's URL and Kuk (1990) with $p(0 < p < 1, p \neq \frac{1}{2})$
as the proportion of A-marked cards

$$y_i = 1 \text{ if } i \text{ bears } A$$
$$= 0 \text{ if } i \text{ bears } A^C, i \in U = (1, \dots, N)$$

and the RR from i as

$$I_i = 1 \text{ if card-type drawn 'matches' } i's \text{ feature}$$
$$= 0 \text{ if card-type drawn 'mismatches' } i's \text{ feature} A \text{ or } A^C.$$

$$r_i = \frac{I_i - (1-p)}{(2p-1)} \text{ has } E_R(r_i) = y_i, V_R(r_i) = V_i$$

equals $\dfrac{p(1-p)}{(2p-1)^2} \forall i \in U.$

Let $L_i \equiv$ the Prob$(y_i = 1)$, $0 < L_i < 1 \forall i$ and be unknown.
Then, posterior probability that

$$L_i(1) = \text{Prob}(y_i = 1| \text{ given } I_i = 1)$$
$$= \frac{L_i P(I_i = 1|y_i = 1)}{L_i P(I_i = 1|y_i = 1) + (1 - L_i)L_i P(I_i = 1|y_i = 0)}$$
$$= \frac{pL_i}{(1 - p) + (2p - 1)L_i} \text{ and}$$

$$L_i(0) = \frac{L_i P(I_i = 0|y_i = 1)}{L_i P(I_i = 0|y_i = 1) + (1 - L_i)L_i P(I_i = 0|y_i = 0)}$$

$$= \frac{(1 - p)L_i}{p + (1 - 2p)L_i}$$

As $p \to \frac{1}{2}$, both $L_i(1) \to L_i$ and $L_i(0) \to L_i$ implying privacy goes on more and more to be protected, but $V_i \to \infty$ i.e., privacy protection and efficiency move inversely as p.

If a response is R and the parameter we intend to estimate is θ, thus $R = 1$ or 0 in Warner's RR, $\theta = \frac{1}{N}\sum_1^N y_i = \frac{Y}{N}$ and $L_i(R)$ is the posterior probability that Prob$(y_i = 1/R)$ then

$$J_i(R) = \frac{L_i(R)/L_i}{(1 - L_i(R))/(1 - L_i)}, i \in U$$

is a 'Measure of jeopardy' in respect of the person labelled i. The closer the value of R_i to unity the less the jeopardy in getting the person's feature be revealed. Nayak (1994) is a pioneering reference to this topic.

More generally,

$$J(R) = \frac{P(A|R)/\theta}{P(A^C|R)/(1 - \theta)}$$

is a 'Measure of jeopardy' if θ is as before and so are A, A^C.

In case θ is as before and so are A and A^C but instead of $I_i = 1$ or $I_i = 0$ more RR's may emerge for an RRI and writing $J_i(R)$ as the $J(R)$ above for a specific person labelled i, then $\bar{J}_i$ is the average of $J_i(R)$ over the different possible RR's from a person i is taken as the 'Average Jeopardy Measure' for a person i. The closer this $\bar{J}_i$ to unity, the greater the privacy protected and the less the chance of revelation of the person's stigmatizing characteristic.

In particular, for Warner (1965) RRT we need to evaluate

$$J_i(1) = \frac{p}{(1 - p)}, J_i(0) = \frac{(1 - p)}{p}$$

and

$$\bar{J}_i = \frac{1}{2}(J_i(1) + J_i(0))$$

$$= \frac{1}{2}\left(\frac{p}{1 - p} + \frac{1 - p}{p}\right).$$

Clearly, $J_i(1) = J_i(0) = \bar{J}_i = 1$ if $p = \frac{1}{2}$. We shall presently tabulate L_i, $L_i(1)$, $J_i(1)$, $J_i(0)$, $\bar{J}_i$, V_i versus various choices of p.

Next we consider (Kuk, 1990 RRT for which L_i, y_i, $(i \in U)$, θ are same as for Warner (1965) the RR's are not confined just to only two namely 1 and 0 but may be more. Here 2 boxes respectively with Red and Non-Red cards in proportions $p_1 : (1 - p_1)$, $(0 < p_1 < 1)$ in the 1-st and $p_2 : (1 - p_2)$, $(0 < p_2 < 1, p_1 \neq p_2)$ are offered to a respondent i in U, and on a request to draw from the 1-st box, if bearing A and from the 2-nd box, if bearing A^C, a fixed number of K draws by SRSWR and report the number f_i of Red balls drawn, $K \geq 2$. Then

$$E_R(f_i) = K[p_1 y_i + p_2(1 - y_i)],$$
$$V_R(f_i) = K[p_1(1 - p_i)y_i + p_2(1 - p_2)(1 - y_i)]$$

For $r_i = \left(\dfrac{f_i}{K}\right) / (p_1 - p_2)$, $E_R(r_i) = y_i$ and

$$V_i = V_R(r_i) = \begin{cases} \dfrac{p_1(1-p_1)}{K(p_1-p_2)^2} & \text{if } y_i = 1 \\ \dfrac{p_2(1-p_2)}{K(p_1-p_2)^2} & \text{if } y_i = 0 \end{cases}$$

$L_i(f_i) \equiv$ the posterior probability that i bears A given the RR form him/her is f_i equals

$$\frac{L_i[p_1^{f_i}(1 - p_1)^{K-f_i}]}{p_2^{f_i}(1 - p_2)^{K-f_i} + L_i[p_1^{f_i}(1 - p_1)^{K-f_i} - p_2^{f_i}(1 - p_2)^{K-f_i}]}$$

As $p_1 \to p_2$ though $L_i(f_i) \to L_i$ yet $V_i \to \infty$

Also $J_i(f_i) \to 1$ if $p_1 \to p_2$ and

$$\bar{J}_i = \frac{1}{(K + 1)} \sum_{f_i=0}^{K} J_i(f_i), \text{ noting}$$

$$J_i(f_i) = \frac{p_1^{f_i}(1 - p_1)^{K-f_i}}{p_2^{f_i}(1 - p_2)^{K-f_i}}$$

So, $\bar{J}_i \to 1$ if $p_1 \to p_2$

Turning thirdly to Simmons's URL, let A be a stigmatizing and B an unrelated innocuous feature,

$$y_i = 1/0 \text{ if } i \text{ bears } A \text{ or } A^C \text{ and}$$
$$x_i = 1/0 \text{ if } i \text{ bears } B \text{ or } B^C.$$

Let a sampled person i be approached with 2 boxes respectively containing A-marked and B-marked cards in proportions $p_1 : (1 - p_1), 0 < p_1 < 1$ and $p_2 : (1 - p_2), 0 < p_2 < 1$ but $p_1 \neq p_2$. On request the person i is to yield RR's using the 1-st box as

$$I_i = 1 \text{ if using 1-st box gets card matching the feature}$$
$$= 0, \text{ else and independently a repeat}$$

I_i' identically distributed as $I_i, i \in U$.
Similarly, independently, using the 2-nd box the RR as

$$t_i = 1 \text{ if the feature matches}$$
$$= 0, \text{ if it does not}$$

and independently another as t_i' distributed identically as t_i. Then,

$$E_R(I_i) = p_1 y_i + (1 - p_1)x_i = E_R(I_1')$$
$$E_R(t_i) = p_2 y_i + (1 - p_2)x_i = E_R(t_1'), i \in U$$

Let $r_i' = \dfrac{(1 - p_2)I_i - (1 - p_1)t_i}{(p_1 - p_2)}$

$$r_i'' = \dfrac{(1 - p_2)I_i' - (1 - p_1)t_i'}{(p_1 - p_2)}$$

$$r_i = \frac{1}{2}(r_i' + r_i''), i \in U.$$

Then, $E_R(r_i') = y_i = E_R(r_i'') = E_R(r_i)$

and $V_i = V_R(r_i) = \dfrac{1}{2(p_1 - p_2)^2} \left[(1 - p_2)^2 V_R(I_i) + (1 - p_1)^2 V_R(t_i) \right]$

$$= \dfrac{(y_i - x_i)^2}{2(p_1 - p_2)^2}(1 - p_1)(1 - p_2)p_1(1 - p_2) + p_2(1 - p_1)$$

The RR's $(I_i, t_i), (I_i', t_i')$ are of the forms $(1, 1), (1, 0), (0, 1)$ and $(0, 0)$. So, the posteriors are also $L_i(1, 1)$ etc. Now,

$$L_i(1, 1) = \dfrac{L_i p_1 p_2}{(1 - p_1)(1 - p_2)} + L_i(p_1 + p_2 - 1)$$
$$\text{and } L_i(1, 1) \to L_i \text{ as } (p_1 + p_2) \to 1$$

But even with $(p_1 + p_2) \to 1$, V_i may yet retain a finite value.
Also, we may define $J_i(1, 1)$ etc and

$$\bar{J}_i = \frac{1}{4} \left[J_i(1, 1) + J_i(1, 0) + J_i(0, 1) + J_i(0, 0) \right]$$

with $J_i(1, 1) = \dfrac{p_1 p_2}{(1 - p_1)(1 - p_2)}$, etc

But $J_i(1, 1) \to 1$ when $(p_1 + p_2) \to 1$

$$\bar{J}_i = \frac{1}{4} \left[\dfrac{p_1 p_2}{(1 - p_1)(1 - p_2)} + \dfrac{p_1(1 - p_2)}{p_2(1 - p_1)} + \dfrac{p_2(1 - p_1)}{p_1(1 - p_2)} + \dfrac{(1 - p_1)(1 - p_2)}{p_1 p_2} \right].$$

So, $\bar{J}_i \to \dfrac{1}{2} + \dfrac{1}{2} \left[\left(\dfrac{p_1}{1 - p_1} \right)^2 + \left(\dfrac{1 - p_1}{p_1} \right)^2 \right]$

as $(p_1 + p_2) \to 1$.

To note that $J_i(1, 0) \to 1$ as $p_1 + p_2 \to 1$, $J_i(0, 1) \to 1$ as $p_1 + p_2 \to 1$ but that $\bar{J}_i \not\to$ unless $p_1 \to p_2$.

But when $p_1 + p_2 \to 1$ and $p_1 \to p_2$ together, it happens that $V_i \to \infty$.

Let us tabulate certain numerical results (Tables 6.1, 6.2 and 6.3).

Next, let us consider more elaborate empirical illustrations considering two of the RRT's due to Warner (1965) and Simmons's (1967, 1969) about their efficiency levels in the estimation as well the levels at which privacy is protected.

Chaudhuri et al. (2009) consider 117 households with given total expenses and classified into those under-reporting Income Taxes (IT), i.e., bearing the stigmatizing feature A and those not doing so, i.e., of the category A^C in respective proportions θ equal to $\frac{82}{117}$. They are also respectively identified into the category B preferring sports to music and the complementary class B^C, with the proportion bearing B as $\frac{85}{117}$.

A sample of size $n = 34$ is drawn following (Rao et al., 1962) sampling scheme employing known size-measures namely the household expenses in a recent month. For this optimal group-sizes are taken so as to optimize the variance of the RHC estimator for the total population households bearing A.

Estimation is studied by simulation of 1000 repeated samples and calculating ACV (Average Coefficient of Variation) ACP(Average Coverage Percentage) for the 95% confidence intervals CI using normal approximation for the standardized error measure, AL (average length of the calculated CI's).

Some findings are tabulated below (Tables 6.4 and 6.5).

Table 6.1 Efficiency and Privacy Protection. Warner's RRT

p	L_i			V_i	$J_i(1)$	$J_i(0)$	$\bar{J}_i$
	$L_i(1)$						
	0.2	0.4	0.7				
0.44	0.164	0.343	0.647	17.1	0.79	1.27	1.03
0.59	0.264	0.489	0.770	7.5	1.44	0.69	1.07

Table 6.2 Kuk's RRT

p		L_i			V_i		$J_i(f_i)$	$\bar{J}_i$
		$L_i(f_i)$						
		0.2	0.4	0.7				
p_1	p_2				$y_i = 1$	$y_i = 0$		
$K = 2,$	$f_i = 0$							
0.52	0.58	0.25	0.47	0.75	34.7	33.8	1.31	1.04
$K = 2,$	$f_i = 1$							
0.52	0.58	0.20	0.41	0.71	34.7	33.8	1.02	1.04

Table 6.3 Simmon's RRT

L_i		$L_i(1, 1)$			V_i	$J_i(1, 1)$	$J_i(1, 0)$	$J_i(0, 1)$	$J_i(0, 0)$	$\bar{J}_i$
p_1	p_2	0.2	0.4	0.7						
0.33	0.51	0.11	0.25	0.54	2.56	0.51	0.47	2.11	1.95	1.26
0.24	0.72	0.17	0.35	0.65	0.28	0.81	0.12	8.14	1.23	2.58

Table 6.4 Warner's RRT (Performances based on simulated samples)

p	$ACV(\%)$	$ACP(\%)$	AL	$J_i(1)$	$J_i(0)$	$\bar{J}_i$
0.60	54.54	97.2	2.41	1.50	0.67	1.08
0.66	54.89	97.4	1.48	1.94	0.52	1.23
0.73	34.83	97.1	1.06	2.70	0.37	1.54
0.75	31.58	98.4	1.00	3.00	0.33	1.67

Table 6.5 Simmon's URL RRT (Performances based on simulated samples)

p_1	p_2	$ACV(\%)$	$ACP(\%)$	AL	$J_i(1, 1)$	$J_i(1, 0)$	$J_i(0, 1)$	$J_i(0, 0)$	$\bar{J}_i$
0.80	0.78	13.38	94.20	0.44	0.69	1.13	0.89	1.46	1.04
0.94	0.84	12.81	92.90	0.43	2.11	2.98	0.34	0.48	1.48
0.75	0.88	12.16	92.30	0.40	0.32	0.41	2.44	3.66	1.58
0.65	0.81	12.94	93.30	0.43	0.29	0.44	2.30	3.50	1.63

References

Chaudhuri, A., & Mukerjee, R. (1988). *Randomized response: Theory and techniques*. N.Y.: Marcel Dekker.

Chaudhuri, A. (2011). *Randomized response and indirect questioning techniques in surveys*. Florida, USA: CRC Press.

Chaudhuri, A., & Christofides, T. C. (2013). *Indirect questioning in sample surveys*. Berlin: Springer.

Chaudhuri, A., Christofides, T. C., & Rao, C. R. (2016). *Handbook of statistics, data gathering, analysis and protection of privacy through randomized response techniques: qualitative and quantitative human traits*, Vol. 34. NL: Elsevier.

Warner, S. L. (1965). RR: A survey technique for eleminating a evasive answer bias. *JASA, 60*, 63–69.

Chaudhuri, A., & Mukerjee, R. (1985). Optionally randomized response techniques. *CSA Bulletin, 34*, 225–229.

Greenberg, B. G., Abul-Ela, A. L., Simmons, W. R., & Gorvito, D. G. (1969). The unrelated question RR model: Theoretical Framework. *JASA, 64*, 520–539.

Chaudhuri, A., & Roy, D. (1997). Model assisted survey sampling strategies with randomized responses. *JSPI, 60*, 61–68.

Chaudhuri, A., & Pal, S. (2015). On efficiency of empirical Bayes estimation of a finite population mean of a sensitive variable through randomized responses. *MASA, 10*, 283–288.

Mangat, N. S. (1992). Two stage randomized response sampling procedure using unrelated question. *JISAS, 44*(1), 82–88.

Liu, P. T., Chow, L. P., & Mosley, W. H. (1975). Use of RR techniques with a new randomizing device. *JASA, 70*, 329–332.

Rao, J. N. K., Hartley, H. O., & Cochran, W. G. (1962). On a simple procedure of unequal probability sampling without replacement. *JRSS B, 24,* 482–491.

Kuk, A. Y. C. (1990). Asking sensitive questions indirectly. *Biometrika, 77*(2), 436–438.

Chaudhuri, A., Christofides, T. C., & Saha, A. (2009). Protection of privacy in efficient application of randomized response techniques. *Statistical Methods and Applications, 18,* 389–418.

Chapter 7
Super-Population Modeling.
Model-Assisted Approach. Asymptotics

7.1 Super-Population: Concept and Relevance

We know $U = (1, \ldots, i, \ldots, N)$ is a finite population of identifiable units assigned respective labels $i = 1, \ldots, N$. On this is defined a vector $\underline{Y} = (y_1, \ldots, y_i, \ldots, y_N)$ of values $y_i, i = 1, \ldots, N$, of a real valued variable y. In our classical design-based theory of survey sampling, the values $y_i, i = 1, \ldots, N$ are supposed to be unknown but fixed numbers. Also, $\underline{Y} = (y_1, \ldots, y_i, \ldots, y_N)$ is supposed to be a specific single element in the space of totality

$$\Omega = \left\{ \underline{Y} \mid -\infty < a_i \le y_i \le b_i < +\infty, i = 1, \ldots, N \right\}$$

with a_i and b_i as real numbers known or unknown. With such a single variable y to start with, we had elaborate theoretical and practical discussions about estimator of parameters

$$Y = \sum_1^N y_i, \; \bar{Y} = \frac{Y}{N}, \; R = \frac{Y}{X} = \frac{\bar{Y}}{\bar{X}}$$

$$R_N = \frac{\sum_1^N (y_i - \bar{Y})(x_i - \bar{X})}{\sqrt{\sum_1^N (y_i - \bar{Y})^2} \sqrt{\sum_1^N (x_i - \bar{X})^2}}, \; \text{etc.}$$

When another real variable x is taken care of with values $x_i, i = 1, \ldots, N$ similarly unknown or known to start with, we consider estimating these finite population parameters $Y, \bar{Y}, R$ in the Chaps. 1–3.

A broad message there emerged that there does not exist among all design-unbiased estimators for Y one with the uniformly least variance

$$E = E_p(t(s, \underline{Y}) - Y)^2$$

© The Author(s), under exclusive license to Springer Nature Singapore Pte Ltd. 2022
A. Chaudhuri and S. Pal, *A Comprehensive Textbook on Sample Surveys*, Indian Statistical Institute Series, https://doi.org/10.1007/978-981-19-1418-8_7

among all unbiased estimators for Y based on whatever design p, the estimator $t = t(s, \underline{Y})$ may be, with $E_p(t) = Y \, \forall \, \underline{Y} \in \Omega$.

So, the classical design-based approach in Survey Sampling is required to be amended.

One such way to effect a change is to regard $\underline{Y}$ as a random vector. Since $\underline{Y}$ is regarded as random, it is supposed to have a probability distribution. Every probability distribution defines a population. The population representing the probability distribution of $\underline{Y}$ is called a 'Super-population' as contrasted with the finite, concrete survey population $U = (1, \ldots, i, \ldots, N)$.

In the context of survey sampling, the probability distribution of $\underline{Y}$ is not required to be too closely specified. It is enough to regard it just as a member of a wide class of probability distributions just admitting a few finite low order moments like mean, variance, covariance, etc.

Such a class of probability distributions of $\underline{Y}$ is called 'Model' for $\underline{Y}$. Writing E_m, V_m, C_m as the operators for mean, variance and covariance computation with respect to a probability distribution suitably modelled we may search for an estimator t for Y so that the 'model-expected' design variance

$$V_p(t) = E_p(t - E_p(t))^2$$
$$= E_p(t - Y)^2 \text{ when } E_p(t) = Y \, \forall, \, \underline{Y}.$$

namely $E_m V_p(t)$ is minimized.

7.2 Godambe-Thompson's (1977) Optimality Approach

Let $t = t(s, Y)$ be a design-unbiased estimator for Y and $t_{HT} = t_{HT}(s, \underline{Y}) = \sum_{i \in s} \frac{y_i}{\pi_i}$

be the Horvitz and Thompson's design-unbiased estimator for $Y = \sum_1^N y_i$.

$$\text{Thus } E_p(t) = \sum_s p(s) t(s, \underline{Y})$$

$$= Y = E_p(t_{HT})$$

$$\text{Let } h(s, \underline{Y}) = t(s, \underline{Y}) - t_{HT}(s, \underline{Y})$$

so that this $h = h(s, \underline{Y})$ is an unbiased estimator for O.

Let $\underline{Y} = (y_1, \ldots, y_i, \ldots, y_N)$ be a random vector such that

$$E_m(y_i) = \mu_i, V_m(y_i) = \sigma_i^2 \text{ and}$$

y_i's be independent of each other for all i's in $U = (1, \ldots, i, \ldots, N)$.

Assuming E_p commutes with E_m, it follows

$$E_m V_p(t) = E_p E_m (t - Y)^2 \tag{7.2.1}$$

$$= E_p E_m \left[(t - E_m(t)) + (E_m(t) - E_m Y) - (Y - E_m Y) \right]^2 \tag{7.2.2}$$

$$= E_p V_m(t) + E_p \Delta_m^2(t) - V_m(Y) \tag{7.2.3}$$

writing $\Delta_m(t) = E_m(t - Y)$ and using definition of t as an estimator.

Since $O = E_p(h) = \sum_{s \not\ni i} p(s)h(s, \underline{Y}) + \sum_{s \ni i} p(s)h(s, Y)$, we may check that

$$E_p V_m(t) = E_p V_m(t_{HT} + h) \tag{7.2.4}$$

$$= E_p V_m(t_{HT}) + E_p V_m(h) \tag{7.2.5}$$

because $E_p C_m(t_{HT}, h) = 0$ as can be checked easily, (vide Chaudhuri (2010)). One may see Chaudhuri (2010) text, p. 79, a companion book of this text.

$$\text{So, } E_p V_m(t) = \sum_1^N \frac{\sigma_i^2}{\pi_i} + E_p V_m(h) \tag{7.2.6}$$

$$\text{and } E_m V_p(t) = \sum_1^N \sigma_i^2 \left(\frac{1}{\pi_i} - 1 \right) + E_p \Delta_m^2(t) + E_p V_m(h) \tag{7.2.7}$$

Now,
$$\Delta_m(t) = E_m(t_{HT}) + E_m(h) - \sum_1^N \mu_i$$

$$= \sum_{i \in s} \frac{\mu_i}{\pi_i} - \sum_1^N \mu_i + E_m(h)$$

Let us choose h as $h_\mu = -\sum_{i \in s} \frac{\mu_i}{\pi_i} + \sum_1^N \mu_i$ and hence t as $t_\mu = t_{HT} + h_\mu = \sum_{i \in s} \left(\frac{y_i - \mu_i}{\pi_i} \right) + \sum_1^N \mu_i$.

Then follows $\Delta_m(t_\mu) = 0$, $V_m(h_\mu) = 0$, $E_p V_m(h_\mu) = 0$

$$\text{So, } E_m V_p(t_\mu) = \sum_1^N \sigma_i^2 \left(\frac{1}{\pi_i} - 1 \right) \text{ and}$$

$$E_m V_p(t) \geq E_m V_p(t_\mu). \text{ Hence, the result.}$$

Under the postulated Model above, $t_\mu = \sum_{i \in s} \frac{y_i - \mu_i}{\pi_i} + \sum_1^N \mu_i$ is the optimal estimator among all design-unbiased estimators t for Y in the sense that

$$E_m V_p(t) \geq E_m V_p(t_\mu) = \sum_1^N \sigma_i^2 \left(\frac{1}{\pi_i} - 1 \right). \tag{7.2.8}$$

We shall often write $MV = \sum_1^N \sigma_i^2 \left(\frac{1}{\pi_i} - 1 \right)$.

The Cauchy Inequality

$$\left(\sum_1^N \pi_i \right) \left(\sum_1^N \frac{\sigma_i^2}{\pi_i} \right) \geq \left(\sum_1^N \sigma_i \right)^2$$

gives equality if $\pi_i \propto \sigma_i$ yielding the minimum value for MV

$$\text{as } \frac{\left(\sum_1^N \sigma_i \right)^2}{\nu} - \sum_1^N \sigma_i^2$$

$$\text{writing } \nu = \sum_1^N \pi_i \equiv \text{ Expected sample-size.}$$

Note 1: If the postulated model be tenable, then an optimal sampling design is one for which $\pi_i = \frac{\nu \sigma_i}{\sum_1^N \sigma_i}$ for $i \in U$ and the optimal design-unbiased estimator is t

equal to $t_\mu = \sum_{i \in s} \left(\frac{y_i - \mu_i}{\pi_i} \right) + \sum_1^N \mu_i$.

In case positive numbers x_i for $i = 1, \ldots, N$ are available and $\mu_i = \beta x_i$ with a certain constant β and $\nu = n$, a positive integer, then the optimal estimator is

$$t_\beta = \sum_{i \in s} \frac{y_i}{\pi_i} + \beta \left(X - \sum_{i \in s} \frac{x_i}{\pi_i} \right) \tag{7.2.9}$$

and the optimal design on which this is to be based should satisfy

$$\pi_i = \frac{n x_i}{X}, i = 1, \ldots, N, X = \sum_1^N x_i.$$

Such a design is called a πPS design or IIPS design which is an 'Inclusion Probability Proportional to size' design with n as the fixed sampling size design.

Note 2: The most important point to observe is that though the above results given by Godambe and Thompson (1977) are mathematically elegant, the estimator t_μ or t_β cannot be used in practice because μ_i or β are not available for use.

A way out, however, is to resort to an 'Asymptotic' approach as briefly described below.

The estimator

$$t_\beta = \sum_{i \in s} \frac{y_i}{\pi_i} + \beta \left(X - \sum_{i \in s} \frac{x_i}{\pi_i} \right) \text{ for } Y \qquad (7.2.10)$$

is motivated by the population model for which we write

$$y_i = \beta x_i + \epsilon_i, i = 1, \ldots, N.$$

Here β is a constant and ϵ_i's are independent random variables each with a zero mean and the model variances are some finite positive numbers. Since it is a regression model, a least squares estimator for β may be taken as

$$b_Q = \frac{\sum_{i \in s} y_i x_i Q_i}{\sum_{i \in s} x_i^2 Q_i} \qquad (7.2.11)$$

with Q_i as some positive constants to be suitably assigned. Thus, t_β may be replaced by

$$t_g = \sum_{i \in s} \frac{y_i}{\pi_i} + b_Q \left(X - \sum_{i \in s} \frac{x_i}{\pi_i} \right) \qquad (7.2.12)$$

which is called a 'generalized regression' (or greg) estimator for Y as first given by Cassel, Särndal and Wretman (CSW, 1976, 1977). This is called a 'Model-Assisted Approach' of deriving an estimator t_g for Y. This is because t_β is suggested by the above postulated model but is modified to t_g and the performance characters of this t_g are developed in terms of its design-based properties. But since t_g is not a linear estimator as it involves

$$\left(\frac{\sum_{i \in s} y_i x_i Q_i}{\sum_{i \in s} x_i^2 Q_i} \right) \left(\sum_{i \in s} \frac{x_i}{\pi_i} \right) = A(s), \text{ say,}$$

the ratio of one random variable to another random variable each involving the random variable s. An exact design-based expectation of $A(s)$ is difficult to work out. But (Brewer, 1979) introduced an asymptotic approach to work out an Asymptotic Design Expectation of such a non-linear function of several statistics.

7.3 Brewer's (1979) Asymptotic Approach

Since (Brewer, 1979) a survey population has a finite size, and the selection without replacement yields a sample necessarily with a finite size as well, an asymptotic approach is not naturally applicable in survey sampling.

In order to get over theoretical problems of analysis, (Brewer, 1979) resorted to the following artifice. He conceived the possibility of a finite population itself re-appearing a number of consecutive times and on each of its appearance it is sampled in an independent manner across the re-appearances. Conceiving the re-appearances of the population an infinite number of times and the sample-selections consecutively executed, an infinite number of times in independent manners, Brewer (1979) could apply concepts of limits and convergence of sequences of statistics and their mathematical expectations. Specifically, he made a clever use of Slutsky's theorem (vide Crammer (1946)) involving limits of functions and rational functions of limits. In particular, he considered Slutsky's result

$$\lim R(u_n, v_n, w_n, \dots)$$
$$= R(\lim u_n, \lim v_n, \lim w_n, \dots)$$

with R as a rational function of sequences u_n, v_n, w_n of real numbers and made use of it hitting upon the concept of "Asymptotic Design—Unbiasedness" to yield the result

$$\lim E_p(t_g) = E_p\left(\sum_{i \in s} \frac{y_i}{\pi_i}\right) + \frac{\lim E_p\left(\sum_{i \in s} \frac{y_i x_i Q_i \pi_i}{\pi_i}\right)}{\lim E_p\left(\sum_{i \in s} \frac{x_i^2 Q_i \pi_i}{\pi_i}\right)}\left(X - E_p\left(\sum_{i \in s} \frac{x_i}{\pi_i}\right)\right)$$
$$= Y \tag{7.3.1}$$

claiming t_g thus is an ADU for Y calling 'ADU' as 'Asymptotically Design Unbiased'. Here, of course

$$\lim E_p\left(\sum_{i \in s} \frac{y_i}{\pi_i} x_i Q_i\right) = \sum_1^N y_i x_i Q_i \text{ and}$$

$$\lim E_p\left(\sum_{i \in s} \frac{x_i^2}{\pi_i} Q_i\right) = \sum_1^N x_i^2 Q_i$$

and $\lim E_p$ denotes limited value of a design-expectation under Brewer's asymptotic theory.

Thus, Brewer's model-Assisted Approach gives us the Greg estimator

$$t_g = \sum_{i \in s} \frac{y_i}{\pi_i} + \frac{\sum_{i \in s} y_i^2 x_i Q_i}{\sum_{i \in s} x_i^2 Q_i}\left(X - \sum_{i \in s} \frac{x_i}{\pi_i}\right) \tag{7.3.2}$$

as an 'Asymptotically Design Unbiased' or ADU estimator for Y.

7.4 Further Analysis Under Model-Assisted Approach

Design-based Mean Square Error (MSE) for the ADU estimator t_g for Y is worked out as follows under the Model-Assisted approach along with an estimator thereof.

Let $e_i = y_i - b_Q x_i$, $i \in U$. Further, let

$$B_Q = \frac{\sum_{i=1}^{N} y_i x_i Q_i \pi_i}{\sum_{i=1}^{N} x_i^2 Q_i \pi_i}, \quad E_i = y_i - B_Q x_i, \quad i \in U.$$

Then, $t_g = \sum_{i \in s} \frac{e_i}{\pi_i} + X b_Q$ and this, for a large sample-size, may be approximated by

$$t_G = \sum_{i \in s} \frac{E_i}{\pi_i} + X B_Q. \tag{7.4.1}$$

Then, for a large sample-size the MSE of t_g about Y may be taken approximately as

$$\begin{aligned}
\mathrm{MSE}(t_g) &\simeq V_p(t_G) = V_p \left(\sum_{i \in s} \frac{E_i}{\pi_i} \right) \\
&= \sum_{i<j}^{N} \sum^{N} \left(\frac{E_i}{\pi_i} - \frac{E_j}{\pi_j} \right)^2 (\pi_i \pi_j - \pi_{ij}) \\
&\quad + \sum_{i=1}^{N} \frac{E_i^2}{\pi_i} \alpha_i
\end{aligned} \tag{7.4.2}$$

writing $\alpha_i = 1 + \frac{1}{\pi_i} \sum_{j \neq i} \pi_{ij} - \sum_{1}^{N} \pi_i$, $i \in U$ following (ChaudhuriandPal, 2002), recalling that this α_i may also be written as

$$\alpha_i = \frac{1}{\pi_i} \sum_{s \ni i} v(s) p(s) - v, \quad i \in U,$$

denoting by $v(s)$, the effective size of a sample s and $v = \sum_s v(s) p(s)$, the expected size of samples and $v = \sum_{1}^{N} \pi_i$.

An estimator for $\mathrm{MSE}(t_g)$ is then taken as

$$m(t_g) = \sum_{\substack{i,j \in s \\ i<j}} \sum \left(\frac{e_i}{\pi_i} - \frac{e_j}{\pi_j} \right)^2 \left(\frac{\pi_i \pi_j - \pi_{ij}}{\pi_{ij}} \right) + \sum_{i \in s} \frac{e_i^2}{\pi_i^2} \alpha_i$$

Mukerjee and Chaudhuri (1990) discussed Asymptotic optimality of double sampling plans for generalized regression estimators.

7.5 Lessons

(i) Given a sample chosen by Rao et al. (1962) scheme, work out a generalized regression estimator for the population total. Derive its measure of error and present an estimator thereof.

Solution: Consider the simple linear regression model through the origin

$$y_i = \beta x_i + \epsilon_i, i \in U = (1, \ldots, i, \ldots, N)$$

The RHC's unbiased estimator for Y is

$$t_{\text{RHC}} = \sum_n y_i \frac{Q_i}{p_i},$$

$$Q_i = p_{i1} + \cdots, + p_{iN_i}, \sum_n Q_i = 1.$$

The greg estimator for Y then is

$$t_{g\text{RHC}} = \sum_n y_i \frac{Q_i}{p_i} + b_R \left(X - \sum_n x_i \frac{Q_i}{p_i} \right),$$

$$\text{and } b_R = \frac{\sum_n y_i x_i R_i \frac{Q_i}{p_i}}{\sum_n x_i^2 R_i \frac{Q_i}{p_i}}$$

with R_i as some positive numbers suitably chosen.

$$\text{Let } B_R = \frac{\sum_1^N y_i x_i R_i}{\sum_1^N x_i^2 R_i},$$

$$E_i = y_i - B_R x_i, i \in U$$

$$\text{and } e_i = y_i - b_R x_i, i \in U.$$

Then, supposing a large sample-size and the model above tenable, a measure of error of $t_{g\text{RHC}}$ about Y may be taken as

$$M\left(t_{g\text{RHC}}\right) = A\sum_{i<j}^{N}\sum^{N} p_i p_j \left(\frac{E_i}{p_i} - \frac{E_j}{p_j}\right)^2$$

where $A = \left(\sum N_i^2 - N\right)/N(N-1)$.

A suitable estimator for it may be taken as

$$m\left(t_{g\text{RHC}}\right) = \frac{\sum_n N_i^2 - N}{N^2 - \sum_n N_i^2} \sum_n \sum_n Q_i Q_j \left(\frac{e_i}{p_i} - \frac{e_j}{p_j}\right)^2.$$

References

Brewer, K. R. W. (1979). A class of robust sampling designs for large scale surveys. *JASA, 74,* 911–915.

Chaudhuri, A. (2010). *Essentials of survey sampling.* New Delhi, India: PHI.

Chaudhuri, A., & Pal, S. (2002). On certain alternative mean square error estimators in complex surveys. *JSPI, 104*(2), 363–375.

Crammer, H. (1946). *Mathematical methods of statistics.* Princeton, NJ: Princeton University Press.

Godambe, V. P., & Thompson, M. E. (1977). Robust near optimal estimation in survey practice. *Bulletin of the International Statistical Institute, 47,* 129–146.

Mukerjee, R., & Chaudhuri, A. (1990). Asymptotic optimality of double sampling plans employing generalized regression estimators. *JSPI, 26,* 173–183.

Rao, J. N. K., Hartley, H. O., & Cochran, W. G. (1962). On a simple procedure of unequal probability sampling without replacement. *JRSS B, 24,* 482–491.

Chapter 8
Prediction Approach: Robustness, Bayesian Methods, Empirical Bayes

8.1 Introduction

Suppose a sample s of size n has been chosen somehow from a population $U = (1, \ldots, i, \ldots, N)$ of size N for which the population total of a real variable y is $Y = \sum_1^N y_i$ and $n < N$.

Writing s^C for the units of U not in the sample and $\sum_{i \in s} y_i$ and $\sum_{j \in s^C} y_j$ sampled and unsampled units of U, we have

$$Y = \sum_1^N y_i = \sum_{i \in s} y_i + \sum_{j \in s^C} y_j. \tag{8.1.1}$$

The first term in the Right Hand Side (RHS) of (8.1.1) is fully known assuming that a survey has been successfully accomplished but the second term therein is unknown. In order that, we may reasonably expect to form an idea of the magnitude of Y from the survey data at hand, we may presume that the co-ordinates of

$$\underline{Y} = (y_1, \ldots, y_i, \ldots, y_N)$$

are mutually inter-related and we may somehow suitably pry into that. For this, we need to suppose that (1) $\underline{Y}$ is a random vector with a probability distribution (2) suitably modelled. If we do so, then

$$Y = \sum_1^N y_i$$

turns out to be a random variable.

© The Author(s), under exclusive license to Springer Nature Singapore Pte Ltd. 2022
A. Chaudhuri and S. Pal, *A Comprehensive Textbook on Sample Surveys*, Indian Statistical Institute Series, https://doi.org/10.1007/978-981-19-1418-8_8

Then, we cannot estimate it. But for a reasonable model for $\underline{Y}$, we may consider the model-based expectation of Y as, say

$$E_m(Y) = \sum_1^N E_m(y_i) = \sum_1^N \mu_i, \text{ say}$$

writing $E_m(y_i) = \mu_i, i \in U$.

An estimator for $\sum_1^N \mu_i$, if available, is regarded as a 'Predictor' for Y. This approach of studying finite survey populations was introduced by Brewer (1963) and later more vigorously developed by Royall (1970). The resulting theory is called Brewer-Royall's theory of survey sampling by the "prediction approach".

8.2   Developing Brewer-Royall's Prediction Approach in Survey Sampling

$$\text{Let } y_i = \beta x_i + \epsilon_i, i \in U \tag{8.2.1}$$

with β as an unknown constant, x_i is the value of a real variable x for the unit i of U and these are often assumed to be known and positive with a total $X = \sum_1^N x_i$. Further, ϵ_i's are independently distributed random variables with means $E_m(\epsilon_i) = 0 \, \forall i$ and variances $V_m(\epsilon_i) = \sigma_i^2, \sigma_i > 0 \, \forall i$ but unknown, but often taken as $\sigma_i^2 = \sigma^2 x_i^g$ with $\sigma > 0$ but unknown and g is an unknown constant such that $0 \le g \le 2$. Thus, $\mu_i = \beta x_i, i \in U$.

$$\text{Since } Y = \sum_{i \in s} y_i + \sum_{j \notin s} y_j$$

$$\text{and } E_m(Y|(s, y_i | i \in s)) = \sum_{i \in s} y_i + \beta \sum_{j \notin s} x_j$$

a predictor for Y should be taken as

$$t = t(s, y_i | i \in s) = \sum_{i \in s} y_i + \left(t - \sum_{i \in s} y_i \right)$$

such that $E_m(t) = E_m(Y|s, y_i, i \in s)$

$$= \sum_{i \in s} y_i + \beta \sum_{j \notin s} x_j \text{ so that}$$

$$t = t(s, y_i | i \in s) = \sum_{i \in s} y_i + \widehat{\beta} \sum_{j \notin s} x_j$$

taking $\widehat{\beta} = \widehat{\beta}(s, y_i | i \in s)$ such that

$$E_m(\widehat{\beta}) = \beta \, \forall \, y_i, i \in s.$$

Then, it will follow that

$$E_m(t | s, y_i, i \in s) \text{ equals } E_m(Y | s, y_i | i \in s)$$

for every y_i for $i \in s$.

Such a $t = t(s, y_i, i \in s)$ is then regarded as a 'Model-unbiased' predictor for Y so long as the model stated in (8.2.1) holds. Thus,

$$E_m(t - Y | s, y_i, i \in s) = 0 \, \forall \, y_i, i \in s$$

when a sample s is at hand.

Now, desirably, among such model-unbiased predictors for Y, a preferred one should satisfy

$$E_m \left((t - Y)^2 | s, y_i, i \in s) \right) = E_m(\widehat{\beta} - \beta)^2 \left(\sum_{j \notin s} x_j \right)^2$$

$$\geq E_m(\widehat{\beta}_0 - \beta)^2 \left(\sum_{j \notin s} x_j \right)^2 \qquad (8.2.2)$$

when $\widehat{\beta}_0 = \widehat{\beta}_0(s, y_i, i \in s)$ satisfies

$$E_m(\widehat{\beta}_0) = \beta, \, \forall s, y_i, i \in s. \qquad (8.2.3)$$

Thus, an optimal model-unbiased predictor for Y is

$$t_0 = t_0(s, y_i, i \in s) = \sum_{i \in s} y_i + \sum_{j \notin s} x_j$$

with β_0 as above in (8.2.2) and (8.2.3).

In order to work out such an optimal 't_0' a 'convenient' way is to restrict $\widehat{\beta}, \widehat{\beta}_0$ among the class

$$\widehat{\beta} = \sum_{i \in s} l_i y_i, \widehat{\beta}_0 = \sum_{i \in s} l_{i_0} y_i \text{ such that}$$

$$E_m(\widehat{\beta}) = \beta, E_m(\widehat{\beta}_0) = \beta \, \forall s, y_i, i \in s.$$

Thus, $\widehat{\beta}_0 \sum_{i \in s} l_i x_i = 1$ and $\sum_{i \in s} l_{i_0} x_i = 1$ (8.2.4)

and $E_m(t - Y)^2 = E_m \left[\sum_{i \in s} l_i(y_i - \beta x_i) \right]^2$

$$\geq E_m(t_0 - Y)^2 = E_m \left[\sum_{i \in s} l_{i_0}(y_i - \beta x_i) \right]^2.$$

So, to work out $t_0 = \sum_{i \in s} y_i + \widehat{\beta}_0 \sum_{j \notin s} x_j$, we need to minimize

$$E_m \left[\sum_{i \in s} l_i(y_i - \beta x_i) \right]^2 = \sum_{i \in s} l_i^2 \sigma_i^2$$

with respect to l_i subjected to

$$\sum_{i \in s} l_i x_i = 1.$$

So, we are to solve, taking a Lagrangian undetermined multiplier

$$0 = \frac{\partial}{\partial l_i} \left[\sum_{i \in s} l_i^2 \sigma_i^2 - \lambda(\sum_{i \in s} l_i x_i - 1) \right]$$

$$= 2l_i \sigma_i^2 - \lambda x_i \text{ so as to get}$$

$$l_i = \frac{\lambda x_i}{2\sigma_i^2}$$

so that

$$1 = \frac{\lambda}{2} \sum_{i \in s} \frac{x_i^2}{\sigma_i^2} \text{ or } \frac{\lambda}{2} = \frac{1}{\sum_{i \in s} \frac{x_i^2}{\sigma_i^2}}$$

so that

$$l_{i_0} = \frac{x_i/\sigma_i^2}{\sum_{i \in s} \frac{x_i^2}{\sigma_i^2}}$$

yielding the optimal model-unbiased predictor for Y as

$$t_0 = \sum_{i \in s} y_i + \left(\frac{\sum_{i \in s} \frac{y_i x_i}{\sigma_i^2}}{\sum_{i \in s} \frac{x_i^2}{\sigma_i^2}} \right) \sum_{j \notin s} x_j.$$

Then, a measure of error for t_0 as a predictor for Y is

$$E_m(t_0 - Y)^2 = \left(\sum_{i \in s} l_{i_0}^2 \sigma_i^2\right)\left(\sum_{j \notin s} x_j\right)^2 + \sum_{i \notin s} \sigma_i^2$$

$$= \left(\frac{1}{\sum_{i \in s} \frac{x_i^2}{\sigma_i^2}}\right)\left(\sum_{i \notin s} x_i\right)^2 + \sum_{i \notin s} \sigma_i^2.$$

An interesting simplification ensues if we treat the special case of the model:

$$\sigma_i^2 = \sigma^2 x_i, i \in U.$$

Then, 't_0' simplifies to

$t_0 = X\dfrac{\bar{y}}{\bar{x}}$, called the 'Ratio predictor' for Y.

Then, $E_m\left(X\dfrac{\bar{y}}{\bar{x}} - Y\right)^2$ equals $\dfrac{N\bar{X}(N - n)\bar{x}_e}{n\bar{x}}\sigma^2$

writing $\bar{x}_e = \dfrac{1}{(N - n)}\sum_{i \notin s} x_i.$

Since $E_m(t_0) = E_m(Y) i e E_m(t_0 - Y) = 0$,

$E_m(t_0 - Y)^2 = V_m(t_0 - Y)$, the model-variance of $(t_0 - Y)$.

In $E_m\left(X\dfrac{\bar{y}}{\bar{x}} - Y\right)^2$, the only unknown element is σ^2. So, a model-unbiased estimator for $E_m(t_0 - Y)^2$ may be derived as follows:

$$e_i = y_i - \widehat{\beta}_0 x_i \text{ with } \widehat{\beta}_0 = \frac{\bar{y}}{\bar{x}}.$$

$$\text{So, } e_i = y_i - \left(\frac{\sum_{i \in s} y_i}{\sum_{i \in s} x_i}\right) x_i$$

$$E_m\sum_{i \in s} \frac{e_i^2}{x_i} = E_m\sum_{i \in s} \frac{1}{x_i}\left[(y_i - \beta x_i) - \left\{\sum_{i \in s}\frac{(y_i - \beta x_i)}{\sum_{i \in s} x_i}\right\} x_i\right]^2$$

$$= \sum_{i \in s} \frac{1}{x_i}\left[\sigma^2 x_i + \frac{x_i^2}{\left(\sum_{i \in s} x_i\right)^2}\sigma^2\left(\sum_{i \in s} x_i\right)\right.$$

$$\left. - 2\sigma^2 \frac{x_i^2}{\sum_{i \in s} x_i}\right]$$

$$= (n - 1)\sigma^2.$$

So, a model-unbiased estimator of σ^2 is

$$\widehat{\sigma}^2 = \frac{1}{(n-1)} \sum_{i \in s} \frac{e_i^2}{x_i}$$

So, a model-unbiased estimator of $E_m(t_0 - Y)^2$ is

$$\left[\frac{1}{(n-1)} \sum_{i \in s} \frac{e_i^2}{x_i}\right] \frac{\bar{X}(1-f)N^2}{n\bar{x}}$$

$$= \frac{N^2(1-f)}{n} \frac{\bar{X}\bar{x}_i}{\bar{x}} \widehat{\sigma}^2$$

writing $f = \frac{n}{N}$, the sampling fraction.

This provides a model-unbiased estimator for the model-based measure of error of the ratio predictor for the population total.

8.3   Robustness Issue (A Short Note)

Postulating the model

$$y_i = \beta x_i + \epsilon_i, i \in U$$

with x_i's known, β unknown and ϵ_i's independent random variables with mean $E_m(\epsilon_i) = 0 \,\forall\, i$ and variances $E_m(\epsilon_i) = \sigma^2 x_i, \sigma > 0$ but otherwise unknown, we found

$$t_0 = \sum_{i \in s} y_i + \left(\sum_{i \notin s} x_i\right) \frac{\sum_{i \in s} y_i}{\sum_{i \in s} x_i}$$

as the model-based and model-unbiased optimal linear predictor for Y.

Suppose the above model does not accurately reflect the relationship among the y's and x's, but it is more feasible to rectify the above as

$$y_i = \alpha + \beta x_i + \epsilon_i, i \in U$$

with $\alpha \neq 0$ but otherwise unknown but everything else is intact.

Consequently, it will follow that

$$E_m(t_0) = \left(n\alpha + \beta \sum_{i \in s} x_i\right) + \left(\sum_{i \notin s} x_i\right) \left[\frac{n\alpha}{\sum_{i \in s} x_i} + \beta\right]$$

$$= \left(\frac{N\bar{X}}{n\bar{x}}\right) n\alpha + \beta X \tag{8.3.1}$$

but $E_m(Y) = N\alpha + \beta X \tag{8.3.2}$

If $\bar{x}$ equals $\bar{X}$, then (8.3.1) matches (8.3.2) and t_0 continues to remain model unbiased for Y when the model is revised with a non-zero α vitiating the originally postulated model.

From this, one may observe that the initial predictor 'to' for Y retains its model-unbiasedness even when a non-zero α appears in the model if a sample is properly chosen for which $\bar{x}$ equals $\bar{X}$, and in such a situation, we say that the optimal estimator under the original model continues to remain robust under the model failure only if that sample be chosen for which the mean $\bar{x}$ equals or is closest to the population mean $\bar{X}$.

However, if the original model is tenable, then an appropriate sampling design may be observed as follows to be quite different.

Under the original model the optimal model-unbiased predictor for Y was seen to be

$$t_0 = X\frac{\bar{y}}{\bar{x}} \text{ and}$$

$$E_m(t_0 - Y)^2 = \frac{N^2(1-f)}{n}\bar{X}\frac{\bar{x}_e}{\bar{x}} = V_m(t_0 - Y).$$

Given N, f and $\bar{X}$, this is the minimum if that sample is chosen for which the sample mean $\bar{x}$ is the largest. Thus, in this case, the sampling design should be one for which the sample mean $\bar{x}$ is the largest of all the $\binom{N}{n}$ sample means $\bar{x}$ for the possible number of samples of size n from the population of size N is taken.

8.4 Bayesian Approach

In the context of a finite population, $U = (1, \ldots, i, \ldots, N)$ with $\underline{Y} = (y_1, \ldots, y_i, \ldots, y_N)$ in $\Omega = \{\underline{Y} | (-\infty < a_i \le y_i \le b_i < +\infty) i \in U\}$ and $d = (s, \underline{y})$, the survey data based on a sample $s = (i_1, \ldots, i_n)$ and observations $\underline{y} = (y_j, j \in s)$ with selection-probability $p(s)$ for a non-informative design p involving no y_j for $j \in U$, the 'likehood' is

$$L_{\underline{Y}}(d) = P_{\underline{Y}}(d) = p(s)I_{\underline{Y}}(d).$$

$$\text{Here } I_{\underline{Y}}(d) = 1 \text{ if } \underline{Y} \in \Omega_d$$

$$= 0 \text{ if } \underline{Y} \notin \Omega_d$$

and $\Omega_d = \underline{Y}|y_i$ for $i \in s$ as observed and $-\infty < a_i \le y_i \le b_i < +\infty$ for $i \notin s$ which thus is the part of Ω that is consistent with the observable data d.

Thus, the likelihood here is flat, it is zero for every $\underline{Y}$ inconsistent with the observable data d and it equals $p(s)$, which is free of $\underline{Y}$ for every $\underline{Y}$ consistent with d. Thus, we cannot discriminate among the $\underline{Y}$'s supported by the data d. So, for none of the unobserved *y*s, it is possible to make any inference on the strength of the likelihood under this most general situation about y's when the design is non-informative.

This topic is reported as above from the discussions by Godambe (1966, 1969) and Basu (1969). But there are two distinct ways to tackle this situation as developed in the Survey Sampling literature.

The first is the Bayesian approach, propagated and championed by Basu (1969, 1971). To Basu (1969, 1971), the above flat likelihood

$$L_Y(d) = p(s)I_Y(d)$$

in Survey Sampling is rather a blessing inducing a simple and easy application of a Bayesian approach to this field.

Just postulate a suitable 'Prior' density, say, $q(\underline{Y})$ for the vector $\underline{Y}$.

Then, combining this with the above simple likelihood $L_Y(d)$, it is simple to derive the posterior for $\underline{Y}$ given d, as

$$q_{\underline{Y}^*}(d) = \frac{q(\underline{Y})L_Y(d)}{\int_\Omega q(\underline{Y})L_Y(d)d_\mu(\underline{Y})}.$$

Here μ is a suitable measure and $\int_\Omega$ denotes an integral with respect to this μ-measure.

So, $q_{\underline{Y}^*}(d) = q_{\underline{Y}}t(d)$, where

$$t(d) = \frac{1}{\int_\Omega q(\underline{Y})d_\mu(\underline{Y})}, \text{ for } \underline{Y} \in \Omega_d \text{ and}$$

$$q_{\underline{Y}^*}(d) = 0 \,\forall\, \underline{Y} \notin \Omega_d.$$

Bayesian inference is formulated through a 'Posterior' density. The Posterior is free of a 'sampling design p'. But this, however, does not imply that a Bayes inference is independent of 'How and What data' are derived to induce an inference based on it. Let us clarify below.

Suppose our intention is to estimate the finite population total Y. Then, a statistic $t = t(d)$ based on survey data d is to be employed to estimate it. Suppose we intend to choose an appropriate $t(d)$ for which the square error 'loss' $(Y - t(d))^2$ is to be suitably controlled. A Bayesian principle then dictates minimizing the Posterior 'Risk', namely the Posterior Expectation of this 'loss' which is

$$E_{q^*}(Y - t(d))^2.$$

A choice of $t(d)$ that minimizes this is

$$t^*(d) = E_{q^*}(Y|d),$$

the posterior expectation of Y, given d.

This $t^*(d) = E_{q^*}(Y|d)$ is called the Bayes Estimator of Y. The quantity

$$E_{q^*}(Y - t^*(d))^2 = E_{q^*}(Y - E_{q^*}(Y|d))^2$$

is called the 'Posterior risk' in estimating Y by the Bayes estimate $t^*(d)$ for the square error loss in estimation.

For the prior q, the expectation

$$E_q \left[E_{q^*}(Y - t^*(d))^2 \right]$$

is called the 'Prior Risk' corresponding to the 'Square Error Loss' in estimating Y by $t(d)$. The Bayes estimator $t^*(d) = E_{q^*}(Y|t(d))$ also minimizes this prior risk in the sense that

$$E_q \left[E_{q^*}(Y - t^*(d))^2 \right] \le E_q \left[E_{q^*}(Y - t(d))^2 \right]$$

for the choice $t^*(d)$ for every $t(d)$. Upto this, nothing has been required to be said about how the survey data d has been gathered or what sampling design p was employed to produce the data d.

Now we may consider the 'Average' 'Prior Risk' with respect to a design p which is

$$E_p \left[E_q (E_{q^*}(Y - t^*(d)))^2 \right]$$

If it is possible to choose a sample s_0 for which the data $d_0 = (s_0, y_i | i \in s)$ may be gathered so that

$$E_q E_{q^*} \left(Y - t^*(d_0) \right)^2 \le E_q E_{q^*} \left(Y - t^*(d) \right)^2$$

for every d other than d_0, then the design p_0 is the optimum design for which

$$p_0(s_0) = 1 \text{ and } p_0(s) = 0$$

for every other sample s not s_0. Thus, the strategy $(p_0, t^*(d_0))$ is 'optimal' with the Bayesian approach on considering 'Square Error Loss'.

But it is easier to theoretically describe than may be implemented.

For a very simple model illustrated below, however, it is possible to derive a simple Bayes estimator in a practicable way.

Suppose a sample s of size N is at hand no matter how yielding a sample mean $\bar{y}$ and let n be the sample size. Suppose for the population of size N, the mean be $\bar{Y}$ which is required to be estimated.

Let us postulate a model for which (i) $\bar{y} = \bar{Y} + \epsilon$ such that $\bar{y}|\bar{Y} \sim N\left(\bar{Y}, \frac{V}{n}\right)$, ie, $\bar{y}$ given $\bar{Y}$ is distributed normally with mean $\widehat{\bar{Y}}$ and variance $\frac{V}{n}$, such that V is the sample-variance $s^2 = \frac{1}{n-1} \sum_{i \in s} (y_i - \bar{y})^2$ and (ii) $\bar{Y} \sim N(\mu, \tau)$, ie, $\bar{Y}$ is distributed normally with a mean μ and an unknown variance τ; (iii) moreover, let $\bar{Y} = \mu + \eta$ so that μ, η are unknown but μ is a constant but η is normally distributed with zero mean and variance τ and ϵ is distributed normally with zero mean and variance $\frac{V}{n}$ and ϵ is distributed independently of η.

Then, we may write $\bar{y} = \bar{Y} + \epsilon = \mu + \epsilon + \eta$ and note

$$\text{(a)} \quad \bar{y} \sim N\left(\mu, \frac{V}{n} + \eta\right) \text{ and}$$

$$\text{(b)} \quad \begin{pmatrix} \bar{y} \\ \bar{Y} \end{pmatrix} \sim N\left(\begin{pmatrix} \mu \\ \mu \end{pmatrix}, \begin{pmatrix} \Sigma_{11} & \Sigma_{12} \\ \Sigma_{21} & \Sigma_{22} \end{pmatrix}\right).$$

$$\Sigma_{11} = \frac{V}{n} + \tau, \sum_{12} = E(\bar{y} - \mu)(\bar{Y} - \mu) = \tau = \sum_{21}$$

$$\Sigma_{22} = E(\bar{Y} - \mu)^2 = \sum_{21} = \sum_{12}$$

Then, the conditional distribution of $\bar{Y}$ given $\bar{y}$ is

$$\bar{Y} | \bar{y} \sim N(\mu + \Sigma_{21} \Sigma_{11}^{-1} (\bar{y} - \mu), \Sigma_{22} - \Sigma_{21} \Sigma_{11}^{-1} \Sigma_{12})$$

following (Anderson, 1958). This simplifies to

$$N\left(\frac{\tau}{\frac{V}{n} + \tau} \bar{y} + \frac{\frac{V}{n}}{\frac{V}{n} + \tau} \mu, \frac{\tau \frac{V}{n}}{\frac{V}{n} + \tau}\right).$$

Thus, for a square error loss, the Bayes estimator $\bar{Y}$ is

$$\bar{Y}_B = \frac{\tau}{\frac{V}{n} + \tau} \bar{y} + \frac{\frac{V}{n}}{\frac{V}{n} + \tau} \mu$$

and its Bayes Risk is

$$\frac{\tau \frac{V}{n}}{\frac{V}{n} + \tau}.$$

Clearly, $\bar{Y}_B$ is a convex combination of $\bar{y}$ and μ, if $\bar{Y}$ is well-estimated by $\bar{y}$, then $\frac{V}{n}$ is relatively less compared to τ so that $\bar{y}$ receives a higher weight rather than μ.

But anyway, $\bar{Y}_B$ is also not amenable to computation because μ and τ are both unknown.

To circumvent this hurdle, let us resort to Empirical Bayes Estimation.

8.5   Empirical Bayes Approach

While postulating a prior in conjunction with a modelled likelihood, a Bayes Estimator for a finite population parameter like a population mean or a population total a Bayes Estimate often is found to involve unknowable parameters rendering it unusable in practice. One way to get rid of this is to work out suitable estimates for these model parameters derived from the survey data and replace the involved parameters by their corresponding estimates in the Bayes estimates. The resulting estimates are called Empirical Bayes estimates.

In the previous Sect. 8.4, for $\bar{Y}$, the Bayes estimate was derived as

$$\bar{Y}_B = \frac{\tau}{\tau + \frac{V}{n}} \bar{y} + \frac{\frac{V}{n}}{\tau + \frac{V}{n}} \mu.$$

As τ and μ are essentially unknown, this is unusable.

Introducing an auxiliary correlated variable x with known values x_i for $i \in U = (1, \ldots, i, \ldots, N)$, with $\bar{X}$ as the population mean $\bar{X} = \frac{1}{N} \sum_{1}^{N} x_i$ one may modify the postulated model by taking $\mu = \beta \bar{X}$, with β as an unknown parameter. Then in the resulting Bayes estimate for $\bar{Y}$, namely

$$\bar{Y}_B = \left(\frac{\tau}{\frac{V}{n} + \tau} \bar{y} \right) + \frac{\frac{V}{n}}{\tau + \frac{V}{n}} \beta \bar{X},$$

β may be replaced by its estimate

$$b = \frac{\sum_{i \in s} y_i x_i / \left(\frac{V}{n} + \tau \right)}{\sum_{i \in s} x_i^2 / \left(\frac{V}{n} + \tau \right)}.$$

Next, noting, under the postulated model $A = \sum_{i \in s} (y_i - b x_i)^2 / \left(\frac{V}{n} + \tau \right)$ is distributed as a χ^2-variable with $(n - 1)$ degrees of freedom, the method of moments may be applied equating A to its degrees of freedom $(n - 1)$. Then, the Newton–Raphson method of iteration may be applied repeatedly starting with 0 as the initial estimate of τ in the resulting equation. Each time a solution yields a value for τ, that may be placed in b and in this way, after a few iterations, a stable value for τ as τ_0 and one for b as estimate for β may be derived as β_0. Then results an Empirical Bayes Estimator (EBE, say) for $\bar{Y}$ as

$$\widehat{\bar{Y}}_{EB} = \left(\frac{\widehat{\tau_0}}{\frac{V}{n} + \widehat{\tau_0}} \bar{y} \right) + \left(\frac{\frac{V}{n}}{\frac{V}{n} + \widehat{\tau_0}} \beta_0 \bar{X} \right).$$

Evaluating a measure for its error $E_{q^*} (\bar{Y} - \widehat{\bar{Y}}_{EB})^2$ is not easy. In Chap. 9, we shall have some discussions on a similar issue.

8.6 Lessons

(i) Starting with a simple model-unbiased predictor for $Y = \sum_1^N y_i$ as

$$t_{opt} = \sum_{i \in s} y_i + \left(\sum_{i \notin s} x_i\right) \frac{\sum_{i \in s} y_i x_i / \sigma_i^2}{\sum_{i \in s} x_i^2 / \sigma_i^2}$$

under the model:

$$y_i = \beta x_i + \epsilon_i, \, i \in U \text{ with}$$
$$E_m(\epsilon_i) = 0 \, \forall i \text{ and } V_m(\epsilon_i) = \sigma_i^2, \, i \in U$$

apply (Brewer, 1979) asymptotic model-assisted approach to derive a practicable version of this.

Solution: Since σ_i is unknown, Brewer (1979) introduced a positive weight $w_i (i \in U)$ to replace t_{opt} by

$$t_w = \sum_{i \in s} y_i + \left(X - \sum_{i \in s} x_i\right) \frac{\sum_{i \in s} y_i x_i w_i}{\sum_{i \in s} x_i^2 w_i}$$

and seeks to choose w_i so that $\lim E_p(t_w)$ may equal Y rendering t_w an ADU predictor for Y. Clearly

$$\lim E_p(t_w) = \sum_1^N y_i \pi_i + \left(X - \sum_1^N x_i \pi_i\right) X \left(\frac{\sum_1^N y_i x_i w_i \pi_i}{\sum_1^N x_i^2 w_i \pi_i}\right)$$

Then, the choice

$$w_i = \frac{1 - \pi_i}{\pi_i x_i} \text{ leads to}$$

$$\lim E_p(t_w) = \sum_1^N y_i \pi_i + \left(\sum_1^N x_i (1 - \pi_i)\right) \frac{\sum_1^N y_i (1 - \pi_i)}{\sum_1^N x_i (1 - \pi_i)}$$
$$= Y.$$

So, t_w with $w_i = \frac{1-\pi_i}{\pi_i x_i}$ is ADU for Y and it is called Brewer's predictor denoted as t_B.

(ii) Work out a suitable variance for Brewer's t_B for large sample-size and suggest an appropriate estimator for this variance.

Noting

$$t_B = \sum_{i \in s} y_i + \left(X - \sum_{i \in s} x_i \right) \frac{\sum_{i \in s} y_i \left(\frac{1-\pi_i}{\pi_i} \right)}{\sum_{i \in s} x_i \left(\frac{1-\pi_i}{\pi_i} \right)}$$

and a greg predictor as

$$t_g = \left(\sum_{i \in s} \frac{y_i}{\pi_i} \right) + \left(X - \sum_{i \in s} \frac{x_i}{\pi_i} \right) \frac{\sum_{i \in s} y_i x_i Q_i}{\sum_{i \in s} x_i^2 Q_i}$$

with Q_i as some positive numbers for every $i \in U$.

Taking $Q_i = \frac{1-\pi_i}{\pi_i x_i}$ in t_g, it is easy to check that t_B equals t_g above.

Thus, since t_B is a greg predictor with $Q_i = \frac{1-\pi_i}{\pi_i x_i}$ and an approximately unbiased variance formula for t_g is available, a variance formula for t_B works out. And hence a formula for an estimator for the variance of t_g follows and hence follows that for Var(t_B).

(iii) Show formally that Brewer's ADU predictor is also a Greg predictor.

Solution: Brewer's predictor for Y is of the form

$$t_w = \sum_{i \in s} y_i + \left(\frac{\sum_{i \in s} y_i x_i w_i}{\sum_{i \in s} x_i^2 w_i} \right) \sum_{i \notin s} x_i$$

Then, $\lim E_p(t_w) = \sum_{i=1}^{N} y_i \pi_i + \left(\frac{\sum_1^N y_i x_i w_i \pi_i}{\sum_1^N x_i^2 w_i \pi_i} \right) \sum_1^N x_i (1 - w_i).$

Then, for the choice $w_i = \frac{1-\pi_i}{\pi_i x_i}$, $\lim E_p(t_w) = Y$. So, the Brewer's predictor is

$$t_B = \left(\sum_{i \in s} y_i \right) + \left[\frac{\sum_{i \in s} y_i \left(\frac{1-\pi_i}{\pi_i} \right)}{\sum_{i \in s} x_i \left(\frac{1-\pi_i}{\pi_i} \right)} \right] \sum_{i \in s} x_i$$

.

A generalized regression (Greg) predictor for Y is

$$t_g = \left(\sum_{i \in s} \frac{y_i}{\pi_i} \right) + \left(X - \sum_{i \in s} \frac{x_i}{\pi_i} \right) \frac{\sum_{i \in s} y_i x_i Q_i}{\sum_{i \in s} x_i^2 Q_i}$$

with $Q_i > 0 \forall i \in U = (1, \ldots, N)$.

Taking $Q_i = \frac{1-\pi_i}{\pi_i x_i}$ and equating the coefficients of y_i and x_i in t_B and t_g, it is easily checked that t_B coincides with t_g for this choice of $Q_i = \frac{1-\pi_i}{\pi_i x_i}$.

(iv) Obtain an estimator for an approximate design variance of t_B.

Solution: Use formulae for $V(t_g)$ and $v(t_g)$ and take thereon $Q_i = \frac{1-\pi_i}{\pi_i x_i}$.

References

Anderson, T. W. (1958). *Introduction to multi-variate statistical analysis*. N.Y., USA: Wiley.

Brewer, K. R. W. (1963). Ratio estimation and finite populations: some results deducible from the assumption of an underlying stochastic process. *Australian Journal of statistics, 5*, 93–105.

Brewer, K. R. W. (1979). A class of robust sampling designs for large scale surveys. *JASA, 74*, 911–915.

Basu, D. (1969). Role of the sufficiency and likelihood principles in sample survey theory. *Sankhyā, 31*, 441–454.

Basu, D. (1971). An essay on the logical foundation of statistical inference. In V. P. Godambe & D. A. Sprout (Eds.) (pp. 203–242). Toronto, Canada: Holt, Rinehart & Winston.

Godambe, V. P. (1966). A new approach to sampling from finite populations II distribution-free sufficiency. *JRSS B, 28*, 320–328.

Godambe, V. P. (1969). Admissibility and Bayes estimation in sampling finite populations-V. *The Annals of Mathematical Statistics, 40*, 672–676.

Royall, R. M. (1970). Finite population sampling theory under certain linear regression models. *Biometrika, 57*, 377–387.

Chapter 9
Small Area Estimation and Developing Small Domain Statistics

9.1 Introduction and Formulation of the Problem

For a finite population $U = (1, \ldots, i, \ldots, N)$ let $U_d(d = 1, \ldots, D)$ be its d-th part or 'domain' so that U_d and $U_{d'}$, for $d \neq d'$, its disjoint parts, $U_d \cap U_{d'} = \Phi$, the empty set and their union $\bigcup\limits_{d=1}^{D} U_d$ is co-extensive with the population U. Let, for $\underline{Y} = (y_1, \ldots, y_i, \ldots, y_N)$, the vector of y-values for a real-variate y and $Y = \sum\limits_{1}^{N} y_i$, be the population total of y_i's and $Y_d = \sum\limits_{i \in U_d} y_i$, the d-th domain totals for $d = 1, \ldots, D$.

Let a sample s of a suitable size n be appropriately chosen from U so as to derive on surveying it, an appropriate estimator for Y. Let, the survey data $(s, y_i | i \in s)$ at hand be required to estimate Y and in addition, the domain totals $Y_d, d = 1, \ldots, D$, with good accuracy-levels as well be simultaneously estimated.

If for a parameter θ related to $\underline{Y}$, $t = t(s, y_i | i \in s)$ be an estimator having a suitable variance-estimator $v(t)$, then its estimated coefficient of variation CV is

$$CV(t) = 100 \frac{\sqrt{v(t)}}{|t|}.$$

Conventionally, "t is"

(i) excellent, if CV $\leq$ 10%,
(ii) good, if 10% < CV $\leq$ 20%,
(iii) tolerable, if 20% < CV $\leq$ 30%,
(iv) unacceptable, if CV > 30%.

If $s_d = s \cap U_d$, be the part of the sample s relevant to the domain U_d and n_d be its size so that $n = \sum\limits_{d=1}^{D} n_d$, then it is often encountered that possibly because of n_d

© The Author(s), under exclusive license to Springer Nature Singapore Pte Ltd. 2022
A. Chaudhuri and S. Pal, *A Comprehensive Textbook on Sample Surveys*, Indian Statistical Institute Series, https://doi.org/10.1007/978-981-19-1418-8_9

being too small, the t_d may have a large CV = CV(t_d), writing $t_d = t(s_d, y_i | i \in s_d)$. Such a t_d may be judged unacceptable under such a circumstance, we say that we have a "Small Area Estimation" (SAE) problem. Only the magnitude of n_d does not decide a Small Area Estimation problem. It is the level of accuracy that is the 'decisive criterion' for an SAE problem in practice.

It is the purpose of a statistician to devise suitable procedures to develop appropriate "Small Domain Statistics" in such circumstances. As a matter of fact the term SAE is appropriate when domains are geographical areas or territorial.

$$\text{Let} \qquad I_{di} = 1, \text{ if } i \in U_d$$
$$= 0, \text{ otherwise;}$$

$$y_{di} = y_i I_{di}, Y_d = \sum_{i=1}^{N} y_{di}.$$

If t_{HT} be the Horvitz-Thompson estimator for Y, namely, $t_{HT} = \sum_{i \in s} \frac{y_i}{\pi_i}$, of course presuming $\pi_i > 0 \, \forall i$, then

$$t_{HT}(d) = \sum_{i \in s} \frac{y_{di}}{\pi_i} = \sum_{i \in s_d} \frac{y_i}{\pi_i}.$$

For this of course,

$$E_p(t_{HT}(t_d)) = Y_d$$
$$\text{and } V_p(t_{HT}(t_d)) = \sum \sum_{i < j} (\pi_i \pi_j - \pi_{ij}) \left(\frac{y_{d_i}}{\pi_i} - \frac{y_{d_j}}{\pi_j} \right)^2$$

and an unbiased estimator for it is

$$v_p(t_{HT}(t_d)) = \sum \sum_{i < j \in s_d} \left(\frac{\pi_i \pi_j - \pi_{ij}}{\pi_{ij}} \right) \left(\frac{y_{d_i}}{\pi_i} - \frac{y_{d_j}}{\pi_j} \right)^2$$

provided $\pi_{ij} > 0 \, \forall i \neq j$.

In case n_d is small, CV$(t_{HT}(t_d))$ often turns out unacceptably large.
Let us consider the greg version of this $t_d = t_{HT}(t_d)$.
For this an appropriate model seems to be

$$y_i = \beta_d x_i + \epsilon_i, i \in U_d$$
$$E_m(\epsilon_i) = 0 \, \forall i, V_m(\epsilon_i) = \sigma_i^2, i \in U_d.$$

Then a Greg predictor for Y_d is

$$t_{g_d} = \sum_{i \in s_d} \frac{y_i}{\pi_i} + \left(X_d - \sum_{i \in s_d} \frac{x_i}{\pi_i} \right) \frac{\sum_{i \in s_d} y_i x_i Q_i}{\sum_{i \in s_d} x_i^2 Q_i}$$

$$= \sum_{i \in s_d} \frac{y_i}{\pi_i} + b_{Q_d} \left(X_d - \sum_{i \in s_d} \frac{x_i}{\pi_i} \right) \quad \text{with } Q_i > 0 \, \forall \, i$$

writing $\quad\quad b_{Q_d} = \dfrac{\sum_{i \in s_d} y_i x_i Q_i}{\sum_{i \in s_d} x_i^2 Q_i}.$

For Q_i's one may take $\frac{1}{x_i^g}, \frac{1}{x_i}, \frac{1}{x_i^2}, \frac{1}{\pi_i x_i}, \frac{1-\pi_i}{\pi_i x_i}$, and g satisfying $0 \le g \le 2, i \in U$.

Letting $B_{Q_d} = \dfrac{\sum_{i \in U_d} y_i x_i Q_i \pi_i}{\sum_{i \in U_d} x_i^2 Q_i \pi_i}$

and $E_i = y_i - B_{Q_d} x_i$ and $e_i = y_i - b_{Q_d} x_i$

and approximate Design variance of t_{g_d} is

$$V_p(t_{g_d}) = \sum_{i<j \in U_d} \sum (\pi_i \pi_j - \pi_{ij}) \left(\frac{E_i I_{d_i}}{\pi_i} - \frac{E_j I_{d_j}}{\pi_j} \right)^2$$

and an estimator for it is

$$v_p(t_{g_d}) = \sum_{i<j \in s_d} \sum \left(\frac{\pi_i \pi_j - \pi_{ij}}{\pi_{ij}} \right) \left(\frac{e_i I_{d_i}}{\pi_i} - \frac{e_j I_{d_j}}{\pi_j} \right)^2$$

If n_d happens to be too small, a very few terms are involved in $v_p(t_{g_d})$ and the estimated CV of t_{g_d} may be unacceptably too big.

9.2 Synthetic Approach for Improved Efficiency

A way out is to artificially enhance effectively the size of the sample used to derive a revised estimator for Y_d by manipulating t_{g_d} and artificially manufacturing an alternative estimator planned to have a reduced CV of it.

A reasonable way is to revise the model as below:

To revise the model let us write

$$y_i = \beta x_i + \epsilon_i, i \in U,$$

with β an unknown constant and ϵ_i's as independent random variables with $E_m(\epsilon_i) = 0 \forall i \in U$ and $V_m(\epsilon_i) = \sigma_i^2, \sigma_i > 0 \forall i \in U$.

Then, β may be estimated by

$$b_Q = \frac{\sum_{i \in s} y_i x_i Q_i}{\sum_{i \in s} x_i^2 Q_i}, \ Q_i > 0 \forall i$$

and Y_d may be estimated by the following synthetic greg predictor

$$t_{Sgd} = \sum_{i \in s_d} \frac{y_i}{\pi_i} + b_Q \left(X_d - \sum_{i \in s_d} \frac{x_i}{\pi_i} \right)$$

This t_{Sgd} is called a 'Synthetic' 'greg' predictor for $Y_d, d = 1, \ldots, D$. This is so called because

(i) b_Q is amalgamative as it involves summing over the units in s_d, aggregated across $d = 1, \ldots, D$ and

(ii) the use of b_Q based on the entire sample s in estimating Y_d is 'artificial', rather forcibly supposing the regression lines of y on x to have a common slope irrespective of the differences among the various domains in respect of y and x-related characteristics.

In order to develop formulae for variances of t_{Sgd} and estimates thereof let us proceed as below.

Let $B_Q = \dfrac{\sum_{i \in U} y_i x_i \pi_i Q_i}{\sum_{i \in U} x_i^2 \pi_i Q_i},$

$E_i = y_i - B_Q x_i, e_i = y_i - b_Q x_i$ and

$$V_p(t_{Sgd}) = \sum_{i<j}^{N} \sum^{N} (\pi_i \pi_j - \pi_{ij}) \left(\frac{E_i I_{d_i}}{\pi_i} - \frac{E_j I_{d_j}}{\pi_j} \right)^2$$

and $v_p(t_{Sgd}) = \displaystyle\sum_{i<j \in s}^{N} \sum^{N} \left(\frac{\pi_i \pi_j - \pi_{ij}}{\pi_{ij}} \right) \left(\frac{e_i I_{d_i}}{\pi_i} - \frac{e_j I_{d_j}}{\pi_j} \right)^2$

The practice of estimating Y_d utilizing y_i for i not only in s_d but outside s_d within s is called 'Borrowing Strength' in the context of developing small domain statistics or Small Area Estimation.

Using 'Synthetic greg estimator' t_{Sgd} for Y_d rather than the pure greg estimator t_{gd} for Y_d is an illustration of the use of 'borrowing strength' justifying by postulating the regression lines for y on x with a common slope β for every domain U_d rather than a separate slope β_d for the respective domain U_d.

As the synthetic greg predictor t_{Sgd} utilizes all the n observed y-values rather than the basic greg predictor t_{gd} which uses only the domain-specific n_d values of y_i and n_d is often much less than n, it is expected that the CV of t_{Sgd} should be less than that of t_{gd} and substantially so if $n_d < n$ for a U_d and s_d.

9.3 An Empirical Approach for a Further Potential Efficacy

Area-level modeling is often described by the popular Fay-Herriot (FH) model (Fay and Herriot 1979), which has been widely used in small area estimation taking into account the sampling design and area auxiliary variables that, in general, are more easily available in practice than unit auxiliary variables. In the Bayesian context and following (Fay and Herriot, 1979), the Bayesian formulation may be presented in the present context.

Thus, let us postulate the following modeling.

Writing t_d as a starting estimator for Y_d with an estimated design variance $V_d (> 0)$, let

$$t_d = Y_d + e_d \text{ with } e_d \sim N(Y_d, V_d)$$
$$\text{so that } t_d | Y_d \sim N(Y_d, V_d)$$

independently for $d = 1, \ldots, D$, $Y_d \sim N(\beta X_d, A)$, $A > 0$ and β as unknown constants and A independent of V_d's, $d = 1, \ldots, D$, it follows that marginally $t_d \sim N(\beta X_d, A + V_d)$ so that from Anderson (1958) we get

$$\begin{pmatrix} t_d \\ Y_d \end{pmatrix} \sim N \left(\begin{pmatrix} \beta X_d \\ \beta X_d \end{pmatrix}, \begin{pmatrix} A + V_d & A \\ A & A \end{pmatrix} \right)$$

Consequently,

$$Y_d | t_d \sim N \left(\beta X_d + \frac{A}{A + V_d}(t_d - \beta X_d), \left(\frac{A V_d}{A + V_d} \right) \right)$$

So,

$$Y_{Bd} = \frac{A}{A + V_d} t_d + \frac{V_d}{A + V_d} \beta X_d$$

is the Bayes estimator for Y_d if we consider a square error loss in this estimation. But this is usable only if β and A are known. But they are not. So, we need to estimate A and β and substituting the estimates of A, β in Y_{Bd} an Empirical Bayes estimator for Y_d needs to be derived.

We may proceed as follows.

Supposing provisionally that A is known β may be estimated, following the 'Least Squares' approved by

$$\widehat{\beta} = \frac{\sum_{d=1}^{D} t_d X_d / (A + V_d)}{\sum_{d=1}^{D} X_d^2 / (A + V_d)}$$

Then, since

$$\sum_{d=1}^{D} (t_d - \widehat{\beta} X_d)^2 / (A + V_d)$$

has the chi-square distribution with $(D - 1)$ degrees of freedom its expectation is $(D - 1)$. So, the 'method of moments' may be applied to estimate A from the equation

$$\sum_{d=1}^{D} (t_d - \widehat{\beta} X_d)^2 / (A + V_d) = (D - 1) \qquad (9.3.1)$$

Starting with an initial trial value A_0 for A and using that in $\widehat{\beta}$ and using A_0 and this transformed $\widehat{\beta}$ successively by iteration in (9.3.1), it is possible to derive two estimators β^* and A^* to be used in $\widehat{Y}_{Bd}$ to derive the empirical Bayes estimator $\widehat{Y}_{EBd}$ for Y_d as

$$\widehat{Y}_{EBd} = \frac{A^*}{A^* + V_d} t_d + \frac{V_d}{A^* + V_d} \beta^* X_d$$

Deriving a formula for a measure of error of $\widehat{Y}_{EBd}$ and of an estimator thereof is too difficult and beyond the scope of covering them in this modest text book.

But we may only quote the not too difficult formulae given by Chaudhuri and Maiti (1994), Prasad and Rao (1990) and Ghosh and Rao (1994) given below

$$V(\widehat{Y}_{EBd}) = \left(\frac{A}{A + V_d}\right) V_d + \left(\frac{V_d}{A + V_d}\right)^2$$

$$\frac{X_d^2}{\sum_{d=1}^{D} \left(\frac{X_d^2}{A + V_d}\right)} + \frac{V_d^2}{(A + V_d)^3} \frac{2}{D^2} \sum_{1}^{D} (A + V_d)^2$$

$$\text{and } \widehat{V}(\widehat{Y}_{EBd}) = \left(\frac{A^*}{A^* + V_d}\right) V_d + \left(\frac{V_d}{A^* + V_d}\right)^2$$

$$\frac{X_d^2}{\sum_{d=1}^{D} \left(\frac{X_d^2}{A^* + V_d}\right)} + \frac{4 V_d^2}{(A^* + V_d)^2} \frac{1}{D^2} \sum_{1}^{D} (A^* + V_d)^2$$

9.4 Strengthening Small Domain Statistics by Borrowing Data Across Time by Kalman Filtering

Suppose a population with little changes re-appears a number of times and so denoted as $U_t = (t_1, \ldots, t_i, \ldots, t_{N_t})$ with size N_t for time-points $0, 1, \ldots, t, \ldots, T$.

Let D domains be time-wise denoted as $U_{d_t}, d = 1, \ldots, D$ and $t = 0, 1, \ldots, T$. Let y_{t_i}, x_{t_i} denote y-and x-variate for units $t_1, \ldots, t_{N_i}$ across $t = 0, 1, \ldots, T$ with domain totals Y_{d_t}, X_{d_t}, the former to be estimated or predicted but the later being known.

Let on time t from the population U_t of size N_t a sample s_t of size n_t be drawn with probability $p(s_t)$ according to a design p_t with inclusion-probabilities $\pi_{t_i} (> 0 \forall t_i$ in $U_t)$ and $\pi_{t_{ij}} (> 0$ for every t_i, t_j in $U_t)$.

Let us postulate a model M_t so as to write

$$y_{ti} = \beta_t x_{ti} + \epsilon_{ti}, t_i \in U_t, t = 0, 1, \ldots, T$$

with β_t as unknown constant and ϵ_{ti}'s as independent random variables with mean $E_m(\epsilon_{ti}) = 0 \forall t_i$ and variances $V_m(\epsilon_{ti}) = \sigma_t^2, t_i \in U_t$ and $\sigma_t > 0 \forall t = 0, 1, \ldots, T$.

The above model specification implying postulation of a common slope in the regression of y on x across the domains induces employing for Y_{dt} the use of the synthetic greg predictor, which is

$$g_{dt} = \sum_{t_i \in s_{dt}} \frac{y_{ti}}{\pi_{ti}} g_{sdt_i}$$

writing $\qquad s_{dt} = s_t \cap U_{dt}, d = 1, \ldots, D$

and $\qquad g_{sdt_i} = I_{dt_i} + \left(X_{dt} - \sum_{t_i \in s_t} \frac{x_{t_i}}{\pi_{t_i}} \right) \dfrac{x_{t_i} Q_{t_i} \pi_{t_i}}{\sum_{t_i \in s_t} x_{t_i}^2 Q_{t_i}}$

$$I_{dt_i} = 1 \text{ if } t_i \in s_{dt}$$
$$= 0 \text{ else}$$

$$b_{Qt} = \frac{\sum_{t_i \in s_t} y_{t_i} x_{t_i} Q_{t_i}}{\sum_{t_i \in s_t} x_{t_i}^2 Q_{t_i}}$$

$$e_{t_i} = y_{t_i} - b_{Qt} x_{t_i}, Q_{t_i} > 0 \forall t_i \in U_t$$

a variance estimator for g_{dt} is

$$v_{dt} = \sum_{\substack{t_i, t_j \in s_t \\ t_i < t_j}} \frac{(\pi_{t_i} \pi_{t_j} - \pi_{t_i t_j})}{\pi_{t_i t_j}} \left(\frac{g_{sdt_i} e_{ti}}{\pi_{t_i}} - \frac{g_{sdt_j} e_{tj}}{\pi_{t_j}} \right)^2$$

taking $\pi_{t_i t_j} > 0 \forall t_i, t_j \in U_t$.

In order to further improve upon g_{dt} in estimating/predicting Y_{dt} it is possible as follows to apply Kalman Filtering (vide Meinhold and Singpurwalla (1983)) using survey data for earlier points of time $t = 0, 1, \ldots, (T - 1)$.

For this first we are to estimate

$$V(b_{Qt}) = V_p E_m(b_{Qt}) + E_p V_m(b_{Qt})$$

$$= \sigma_t^2 E_p \left[\frac{\sum_{s_t} x_{t_i}^2 Q_{t_i}^2}{\left(\sum_{s_t} x_{t_i}^2 Q_{t_i} \right)^2} \right]$$

Taking

$$\widehat{\sigma}_t^2 = \frac{1}{(n_t - 1)} \sum_{t_i \in s_t} e_{t_i}^2$$

it follows, vide (Chaudhuri and Maiti, 1994) that

$$W_t = \widehat{\sigma}_t^2 \left[\frac{\sum_{s_t} x_{t_i}^2 Q_{t_i}^2}{\left(\sum_{s_t} x_{t_i}^2 Q_{t_i} \right)^2} \right]$$

is an over-estimate of $V(b_{Qt})$ still, in the absence of a better one we decide to take this W_t to estimate $V(b_{Qt})$.

Now to apply Kalman Filtering let us introduce the following recursive model M_t so as to write

(i) $g_{dt} = \beta_t X_{dt} + e_{dt}$
(ii) $\beta_t = \beta_{t-1} + w_t, t = 1, 2, \ldots, T$
(iii) e_{dt}'s distributed as $N(0, v_{dt})$ independent of each other and also independently of w_t's which are themselves distributed as $N(0, W_t)$ independently of each other
(iv) also β_0 is a random variable distributed as $N(p_{d0}, \sum_{d0})$ with $p_{d0} = \frac{g_{d0}}{X_{d0}}$ and $\sum_{d0} = \frac{v_{d0}}{X_{d0}}$, supposing v_{d0} as strictly positive.

Now from Chaudhuri and Maiti (1994) we may write,

$$\underline{g}_{d(t-1)} = \left(g_{d1}, \ldots, g_{d(t-1)} \right)$$

$$\underline{g}_{dt} = \left(\underline{g}_{d(t-1)}, g_{dt} \right), t = 2, 3, \ldots, T$$

Moreover, let

$$R_{d1} = \sum_{d0} + W_1$$

$$g_{d1}^* = X_{d1} \phi_{d0}, \Delta_{d1} = g_{d1} - g_{d1}^*$$

and writing throughout $\sim$ to mean "distributed as":

$$\Delta_{d1}/g_{d0} \sim N(0, X_{d1}^2 R_{d1} + v_{d1})$$

$$\beta_1/g_{d1} \sim N\left(\phi_{d0} + \frac{R_{d1} X_{d1}}{X_{d1}^2 R_{d1} + v_{d1}} \Delta_{d1}, R_{d1} - \frac{R_{d1}^2 X_{d1}^2}{X_{d1}^2 R_{d1} + v_{d1}}\right)$$

Hence Kalman Filtering is adopted recursively, vide (Meinhold and Singpurwalla, 1983) as follows to derive

$$\beta_{t-1}/\underline{g}_{d(t-1)} \sim N\left(\phi_{d(t-1)}, \sum_{d(t-1)}\right),$$

$$\beta_1/\underline{g}_{d(t-1)} \sim N\left(\phi_{d(t-1)}, R_{d1}\right),$$

Let us write $R_{dt} = \sum_{d(t-1)} + W_t$.

Then, $\Delta_{dt}/\left(\underline{g}_{d(t-1)}, \beta_t\right) \sim N\left(\beta_t - \phi_{d(t-1)}, v_{dt}\right)$

$$\beta_t/\underline{g}_{dt} \sim N\left(\phi_{d(t-1)} + \frac{R_{dt} X_{dt}}{X_{dt}^2 + v_{dt}} \Delta_{dt}, \right.$$

$$\left. R_{dt} - \frac{R_{dt}^2 X_{dt}^2}{X_{dt}^2 R_{dt} + v_{dt}}\right)$$

So, $\widehat{\beta}_t = \phi_{d(t-1)} + \frac{R_{dt} X_{dt}}{X_{dt}^2 R_{dt} + v_{dt}} \Delta_{dt}$

$$= \phi_{dt}, \text{ say}$$

is the Kalman filter estimator of β_t and

$$\widehat{Y}_{dt} = \widehat{\beta}_t X_{dt}$$

is the Kalman estimator of Y_{dt} using data for time points $t = 0, 1, \ldots, t - 1$.
The measure of error of $\widehat{Y}_{dt}$ as an estimator of Y_{dt} is taken as

$$\sum_{dt} X_{dt}^2 \text{ with } \sum_{dt} = R_{dt} - \frac{R_{dt}^2 X_{dt}^2}{X_{dt}^2 R_{dt} + v_{dt}}.$$

It is of interest to examine I (g_{dt}, v_{dt}) Versus II $(\widehat{\beta}_t X_{dt}, \sum_{dt}^2 X_{dt}^2)$.

Though II is derived with a considerable labour vis-a-vis I there is no guarantee that II may be more efficacious than I.

In practice through observed data by dint of simulation one may study II vs I competetive.

The literature, vide (Chaudhuri and Maiti, 1994, 1997; Chaudhuri et al., 1997; Chaudhuri, 2012) illustrate many details of required exercises in this context.

9.5 Discussion

The accumulated research vide the books and monographs by Mukhopadhyay (1998), Rao (2003) and Rao and Molina (2015) and the papers cited therein especially by Malay Ghosh, Gourishankar Datta, Partha Lahiri and Pfefferman among many others is a gigantic volume of huge materials of interest. They mostly explore theoretical results derived under chronologically developed and refined models.

But mostly they are silent about what transpires on starting with an initial estimator based on a varying probability sampling design followed by its amendment by synthetic generalised regression technique followed by (1) empirical Bayesian and (2) Kalman Filtering techniques. SAE in this text is written by the senior author Arijit Chaudhuri. So, let me discuss a little more about it below.

My specific personal approach is amply clarified in my book (vide Chaudhuri (2012)), which is hardly reviewed by the celebrated experts in this field.

Finally, Hierarchical Bayes procedures are also developed by many to cover the SAE but the present author has no contribution to this and hence chooses to remain reticent here.

The SAE has enormously advanced in the intervening years but his emphasis on certain specific aspects of SAE is well-known to the workers in this area. The most prominent books on SAE by Rao (2003) and Rao and Molina (2015) bear testimony to this claim. So, what is presented in the present text may be duly accepted as a relevant material worthy of attention in spite of its limited coverage. Chaudhuri (2012) has discussed the basics of small area estimation theories in a simple way.

9.6 Lessons

(i) If a population total is to be estimated by Rao et al. (1962) procedure, discuss in details how you may employ the synthetic regression estimators for the domain totals.

Recalling that RHC sampling randomly divides the population into n disjoint groups, denote by $p_i = \frac{x_i}{X}$, the normed size-measures, let for the i-th group ($i = 1, \ldots, n$) the N_i units in it have $R_i = \sum_n p_i$ be the summed normed size-measures in it. Then for π_i in case the Horvitz-Thompson estimator might be employed, take the quantity $\frac{p_i}{R_i}$ for $i = 1, \ldots, n$. Then for the model

$$y_i = \beta x_i + \epsilon_i, i \in U$$

take

$$b_{Qd} = \frac{\sum_n y_i x_i Q_i}{\sum_n x_i^2 Q_i}, \qquad b_Q = \frac{\sum_n y_i x_i Q_i}{\sum_n x_i^2 Q_i}$$

$$g_{sd} = \sum_n y_i \frac{R_i}{p_i} I_{di} + b_{Qd} \left(X_d - \sum_n x_i \frac{R_i}{p_i} I_{di} \right)$$

$$B_Q = \frac{\sum_n y_i x_i \frac{R_i}{p_i} Q_i}{\sum_n x_i^2 \frac{R_i}{p_i} Q_i}$$

$$e_i = y_i - b_Q x_i, \, i \in U = (1, \dots, N).$$

Then mimic the formulae when you might initiate with HTE. First recall that the initial estimator by RHC method is

$$\widehat{Y}_d = \sum_n y_{di} \frac{R_i}{p_i},$$

$$V(\widehat{Y}_d) = \frac{\sum_n N_i^2 - N}{N(N-1)} \sum_n \sum_n p_i p_j \left(\frac{y_{di}}{p_i} - \frac{y_{dj}}{p_j} \right)^2$$

$$v(\widehat{Y}_d) = \frac{\sum_n N_i^2 - N}{N^2 - \sum_n N_i^2} \sum_n \sum_n R_i R_j \left(\frac{y_{di}}{p_i} - \frac{y_{dj}}{p_j} \right)^2.$$

The synthetic greg versions of these are easy to work out. Try yourself.

References

Anderson, T. W. (1958). *Introduction to multi-variate statistical analysis*. N.Y., USA: Wiley.

Chaudhuri, A. (2012). *Developing small domain statistics - modeling in survey sampling (e-book)*. Saarbrucken, Germany: Lambert Academic Publishing.

Chaudhuri, A., & Maiti, T. (1994). Borrowing strength from past data in small domain prediction by Kalman filtering—a case study. *Communications in Statistics-Theory and Methods, 23*, 3507–3514.

Chaudhuri, A., & Maiti, T. (1997). Small domain estimation by borrowing strength across time and domain—a case study. *Statistical Computation and Simulation, 26*(4), 1547–1557.

Chaudhuri, A., Adhikary, A. K., & Seal, A. K. (1997). Small domain estimation by empirical Bayes and Kalman filtering procedures—a case study. *Communications in Statistics-Theory and Methods, 26*(7), 1613–1621.

Fay, R. E., & Herriot, R. A. (1979). Estimation of income from small places: An application of James-Stein procedures to census data. *JASA, 74*, 269–277.

Ghosh, M., & Rao, J. N. K. (1994). Small area estimation: An appraisal. *Statistical Science, 81*, 1058–1062.

Meinhold, R. J., & Singpurwalla, N. D. (1983). Understanding the Kalman filter. *The American Statistician, 37*, 123–127.

Mukhopadhyay, P. (1998). *Theory and methods of survey sampling*. Prentice Hall of India.

Prasad, N. G. N., & Rao, J. N. K. (1990). The estimation of the mean squared error of small area estimations. *JASA, 85*, 163–171.

Rao, J. N. K. (2003). *Small area estimation*. NY, USA: Wiley Interscience.

Rao, J. N. K., & Molina, J. (2015). *Small area estimation* (2nd ed.). N.Y., USA: Wiley.

Rao, J. N. K., Hartley, H. O., & Cochran, W. G. (1962). On a simple procedure of unequal probability sampling without replacement. *JRSS B, 24*, 482–491.

Chapter 10
Estimation of Non-linear Parametric Functions

10.1 Linearization

Let $\theta_1, \theta_2, \ldots, \theta_K$ be K population parameters which are linear in y's. We are going to estimate the parameter $f(\underset{\sim}{\theta})$ which is a nonlinear function (f) of θ_i's, $i = 1, 2, \ldots, K$.

So $f(\underset{\sim}{\theta}) = f(\theta_1, \theta_2, \ldots, \theta_K)$.

Defining the linear estimators of θ_i's as t_i, estimator of $f(\underset{\sim}{\theta})$ can be written as $f(\underset{\sim}{t}) = f(t_1, t_2, \ldots, t_K)$.

Expanding $f(\underset{\sim}{t})$ about $f(\underset{\sim}{\theta})$, and using Taylor Series expansion for large n and ignoring the higher order terms, the term $f(\underset{\sim}{t})$ can be written as

$$f(t_1, \ldots, t_K) = f(\theta_1, \ldots, \theta_K) + \sum_{i=1}^{K} \lambda_i (t_i - \theta_i)$$

$$\text{where } \lambda_i = \frac{\delta}{\delta t_i} f(t_1, \ldots, t_K)|_{\underset{\sim}{t} = \underset{\sim}{\theta}}.$$

The variance of $f(t_1, \ldots, t_K)$ can be approximated by the variance of the term $\sum_{i=1}^{K} \lambda_i t_i$ noting λ_i and θ_i as constants.

$$\text{So } V[f(t_1, \ldots, t_K)] \simeq V(\sum_{j=1}^{K} \lambda_j t_j)$$

Writing the linear estimator t_j as $t_j = \sum_{i \in s} b_{s_i} \xi_{ji}$, we may express $V[f(t_1, t_2, \ldots, t_K)]$ as

$$V[f(t_1, \ldots, t_K)] \simeq V(\sum_{j=1}^{K} \lambda_j t_j)$$

$$= V\left[\sum_{j=1}^{K} \lambda_j \left(\sum_{i \in s} b_{s_i} \xi_{ji}\right)\right]$$

$$= V\left[\sum_{i \in s} b_{s_i} \phi_i\right]$$

where b_{s_i} is the sample based weight and $\phi_i = \sum_{j=1}^{K} \lambda_j \xi_{ji}$.

So the variance of the non-linear statistic $f(t_1, \ldots, t_K)$ can be approximated as the variance of the linear function. Adopting HT (Horvitz and Thompson, 1952) method of estimation, the Yates-Grundy (Y-G) form of the variance can be written as

$$V_{YG}[f(t_1, \ldots, t_K)] = V\left(\sum_{i \in s} \frac{\phi_i}{\pi_i}\right)$$

$$= \sum\sum_{i<j=1}^{N} (\pi_i \pi_j - \pi_{ij}) \left(\frac{\phi_i}{\pi_i} - \frac{\phi_j}{\pi_j}\right)^2$$

with $\pi_i > 0 \forall i$. Chaudhuri (2010) narrated the details in his monograph.

As an example we may consider the estimation of the ratio $R = \frac{Y}{X} = \frac{\sum_{1}^{N} y_i}{\sum_{1}^{N} x_i}$ which

is a nonlinear function of two totals Y and X.

So $\qquad K = 2, \theta_1 = \sum_{1}^{N} y_i = Y, \theta_2 = \sum_{1}^{N} x_i = X$ and

$$f(\underline{\theta}) = f(\theta_1, \theta_2) = \frac{\theta_1}{\theta_2} = \frac{Y}{X} = R.$$

Drawing sample with a general sampling design p and employing Horvitz-Thompson (HT, 1952) method of estimation, the related estimators of θ_1 and θ_2 are $t_1 = \sum_{i \in s} \frac{y_i}{\pi_i}$ and $t_2 = \sum_{i \in s} \frac{x_i}{\pi_i}$ with $\pi_i > 0$.

The estimator $f(t_1, t_2)$ of $f(\theta_1, \theta_2)$ is $\widehat{R} = \frac{t_1}{t_2} = f(t_1, t_2)$.

The variance is
$$V(\widehat{R}) = V\left(\frac{t_1}{t_2}\right) \simeq V(\lambda_1 t_1 + \lambda_2 t_2)$$
where
$$\lambda_1 = \frac{\delta}{\delta t_1} f(t_1, t_2)|_{t=\theta} \text{ and } \lambda_2 = \frac{\delta}{\delta t_2} f(t_1, t_2)|_{t=\theta}$$
or,
$$V\left(\frac{t_1}{t_2}\right) \simeq V\left(\sum_{i \in s} \frac{\phi_i}{\pi_i}\right)$$
where
$$\phi_i = \sum_{j=1}^{K} \lambda_j \xi_{ji}.$$

If the sample is drawn by SRSWOR then,

$$t_1 = N\bar{y} = N\left(\frac{\sum_{i \in s} y_i}{n}\right) \text{ and } t_2 = N\left(\frac{\sum_{i \in s} x_i}{n}\right)$$

So,
$$f(t_1, t_2) = \frac{\bar{y}}{\bar{x}} = \widehat{R}$$

$$\lambda_1 = \frac{1}{\bar{X}}$$

and
$$\lambda_2 = \frac{-R}{\bar{X}}$$

Then
$$V(f(t_1, t_2)) = V\left(\frac{\bar{y}}{\bar{x}}\right)$$

$$= V\left[\frac{N}{n} \sum_{i \in s} \frac{1}{\bar{X}} y_i - \frac{R}{\bar{X}} x_i\right]$$

$$= \left(\frac{N}{n} \frac{1}{\bar{X}}\right)^2 V\left[\sum_{i \in s} (y_i - R x_i)\right]$$

$$= \frac{N^2}{\bar{X}^2} \frac{1-f}{n} \frac{1}{N-1} \sum_{i=1}^{N} (y_i - R x_i)^2$$

It may be estimated by

$$\frac{N^2}{\bar{x}^2} \left(\frac{1-f}{n}\right) \frac{1}{n-1} \sum_{i \in s} (y_i - \widehat{R} x_i)^2.$$

The non-linear parameters $R_N(x, y)$, the total correlation coefficient between y and x and $B_N(y|x)$, the total regression coefficient of y on x can be estimated by the above discussed linearization method.

$R_N(x, y)$ is a non-linear function of 6 totals, say, $\theta_1 = N, \theta_2 = \sum_{i=1}^{N} x_i y_i, \theta_3 = \sum_{1}^{N} x_i, \theta_4 = \sum_{1}^{N} y_i, \theta_5 = \sum_{1}^{N} x_i^2$ and $\theta_6 = \sum_{1}^{N} y_i^2$.

$$R_N(x, y) = \frac{N \sum_1^N x_i y_i - \left(\sum_1^N x_i\right)\left(\sum_1^N y_i\right)}{\sqrt{N \sum_{i=1}^N x_i^2 - \left(\sum_{i=1}^N x_i\right)^2}\sqrt{N \sum_{i=1}^N y_i^2 - \left(\sum_{i=1}^N y_i\right)^2}}$$

$$= \frac{\theta_1\theta_2 - \theta_3\theta_4}{\sqrt{\theta_1\theta_5 - \theta_3^2}\sqrt{\theta_1\theta_6 - \theta_4^2}}.$$

Similarly

$$B_N(x, y) = \frac{N \sum_{i=1}^N x_i y_i - \left(\sum_{i=1}^N x_i\right)\left(\sum_{i=1}^N y_i\right)}{N \left(\sum_{i=1}^N x_i^2\right) - \left(\sum_{i=1}^N x_i\right)^2}$$

Usual estimates of $R_N(x, y)$ based on the initial sample s (drawn by any general sampling design) is

$$r_{xy} = \frac{\left(\sum_{i\in s}\frac{1}{\pi_i}\right)\left(\sum_{i\in s}\frac{y_i x_i}{\pi_i}\right) - \left(\sum_{i\in s}\frac{y_i}{\pi_i}\right)\left(\sum_{i\in s}\frac{x_i}{\pi_i}\right)}{\left[\left(\sum_{i\in s}\frac{1}{\pi_i}\right)\left(\sum_{i\in s}\frac{y_i^2}{\pi_i}\right) - \left(\sum_{i\in s}\frac{y_i}{\pi_i}\right)^2\right]^{\frac{1}{2}}\left[\left(\sum_{i\in s}\frac{1}{\pi_i}\right)\left(\sum_{i\in s}\frac{x_i^2}{\pi_i}\right) - \left(\sum_{i\in s}\frac{x_i}{\pi_i}\right)^2\right]^{\frac{1}{2}}}$$

with HT (1952) method of estimation ($\pi_i > 0$)

$$= \frac{t_1 t_2 - t_3 t_4}{\sqrt{t_1 t_5 - t_3^2}\sqrt{t_1 t_6 - t_4^2}}.$$

Writing $t_i = \sum_{i\in s}\frac{\xi_{ji}}{\pi_i}$, $j = 1, 2, \ldots, 6$ Here, $\xi_{1i} = 1, \xi_{2i} = y_i x_i, \xi_{3i} = y_i, \xi_{4i} = x_i, \xi_{5i} = y_i^2$ and $\xi_{6i} = x_i^2$ with $K = 6$.

Now $\lambda_j = \frac{\delta}{\delta t_j} f(t_1, \ldots, t_6)|_{t=\theta}$ will be calculated for all j ($j = 1, 2, \ldots, 6$).

Then $V(r_{xy}) \simeq V\left(\sum_{i\in s}\frac{\phi_i}{\pi_i}\right)$ where $\phi_i = \sum_{j=1}^K \lambda_j \xi_{ji}$ It's the (Horvitz and Thompson, 1952) variance form.

For $B_N(x, y)$ estimation also, we will proceed in the similar manner.

Pal and Shaw (2021) demonstrated the estimation of finite population quantiles and their mean square error with linearization method.

The distribution function of y can be taken as $F(a)$, where

$$F(a) = \frac{1}{N}\sum_{i=1}^N w_i \text{ with } w_i = \begin{cases} 1 & \text{if } y_i \leq a \\ 0, & \text{otherwise.} \end{cases}$$

Drawing a sample s of size n from the population with an unequal probability sampling design, and adopting a suitable model, $F(a)$ can be estimated by a greg estimator $\widehat{F}(a)$ which is a nonlinear function of several linear estimators (Chaudhuri, 2018).

$$\widehat{F}(a) = \frac{1}{N}\left[\sum_{i \in s}\frac{w_i}{\pi_i} + \frac{\sum_{i \in s}w_i x_i Q_i}{\sum x_i^2 Q_i}\left(X - \sum_{i \in s}\frac{x_i}{\pi_i}\right)\right]$$

where $X = \sum_{i=1}^{N} x_i$, Q_i's are positive constants free of $y_1, y_2, \ldots, y_N$.

Writing $\widehat{F}(a)$ as a nonlinear function of several linear estimators, Pal and Shaw (2021) carried out the estimation of $F(a)$ and the related quartile using Linearization technique.

10.2 Jackknife

Jackknife technique was first introduced by Quenouille (1949) as a bias reduction tool. Later (Tukey, 1958) used this technique in estimating mean square errors of biased estimators. So the main application for the Jackknife is to reduce bias of an estimator for moderate sample size.

Let θ be a parameter. A sample s of size n is drawn from the population and $t = t(n)$ be an estimator for θ based on s. The bias B_n is

$$\begin{aligned}
B_n = B_n(\theta) &= E(t(n) - \theta) \\
&= E(t(n)) - \theta \\
&= \frac{b_1(\theta)}{n} + \frac{b_2(\theta)}{n^2} + \frac{b_3(\theta)}{n^3} + \cdots
\end{aligned}$$

where $b_j(\theta)$, $j = 1, 2, \ldots$ are unknown functions of θ and $b_1(\theta) \neq \theta$. In this section, an alternative estimator for θ can be derived with a bias of order $\frac{1}{n^2}$ which is of lower order than the bias of $t(n)$.

The sample s of size n is divided into g disjoint parts each of a size $m = \frac{n}{g}$ such that n, m, g are positive integers.

A new statistic for each group is calculated based on the units of other $(g - 1)$ groups omitting the values of that group. Let for ith group $(1, 2, \ldots, g)$ calculate $t(n - m)$ which is calculated based on $(n - m)$ units omitting the m units of ith group.

A new estimator e_i is calculated as

$$e_i = gt(n) - (g - 1)t_i(n - m).$$

The statistic e_i is termed as pseudo value.

Here
$$
\begin{aligned}
E(e_i) &= E[gt(n) - (g-1)t_i(n-m)] \\
&= gE(t(n)) - (g-1)E(t_i(n-m)) \\
&= g\left[\theta + \frac{b_1(\theta)}{n} + \frac{b_2(\theta)}{n^2} + \cdots\right] \\
&\quad - (g-1)\left[\theta + \frac{b_1(\theta)}{n-m} + \frac{b_2(\theta)}{(n-m)^2} + \cdots\right] \\
&= \theta + b_1(\theta)\left[\frac{g}{n} - \frac{g-1}{n - \frac{n}{g}}\right] \\
&\quad + b_2(\theta)\left[\frac{g}{n^2} - \frac{(g-1)}{\left(n - \frac{n}{g}\right)^2}\right] + \cdots \\
&= \theta - \frac{b_2(\theta)}{n^2}\cdot\left(\frac{g}{g-1}\right) + \cdots
\end{aligned}
$$

For g no of groups, g such pseudo values $e_i (i = 1, 2, \ldots, g)$ can be computed, each with a bias of order $\frac{1}{n^2}$.

The average of these $e_i (i = 1, 2, \ldots, g)$

$$
t_J = \frac{1}{g}\sum_{i=1}^{g} e_i = g(t(n)) - \frac{g-1}{g}\sum_{i=1}^{g} t_i(n-m)
$$

is called a Jackknife statistic with a bias of order $\frac{1}{n^2}$ of the parameter θ.

Starting with t_J and proceeding as above, another estimator with a bias of order $\frac{1}{n^3}$ can be obtained.

The mean square error (MSE) of the Jackknife statistic t_j can be estimated by

$$
\widehat{V}(t_j) = v(j) = \frac{1}{g(g-1)}\sum_{i=1}^{g}\left(e_i - \frac{1}{g}\sum_{i=1}^{g} e_i\right)^2
$$

$$
== \frac{1}{g(g-1)}\sum_{i=1}^{g}(e_i - t_j)^2
$$

The above MSE estimator $v(j)$ may be used to estimate the MSE of non-linear statistics.

To construct $100(1 - \alpha)\%$ confidence intervals for θ for large n, the pivotal $\frac{t_J - \theta}{\sqrt{v(j)}}$ is computed. This pivotal follows student's t distribution with $(g - 1)$ degrees of freedom (df). Then $t_J \pm t_{g-1}, \frac{\alpha}{2}\sqrt{v(j)}$ gives us the $100(1 - \alpha)\%$ confidence intervals for θ. For very large g, the interval becomes $t_J \pm \eta_{\alpha/2}\sqrt{v(j)}$ approximating by that of the standard normal deviate η.

The term t_{g-1}, $\frac{\alpha}{2}$ denotes the $100\frac{\alpha}{2}\%$ point in the right tail of student t statistic with $(g - 1)$ df.

Berger and Skinner (2005) proposed a jackknife variance estimator for any general without-replacement unequal probability sampling design.

Let $f(f_1, \ldots, f(t_K)) = f(\underline{t})$ be the estimator of the non-linear parameter $f(\theta_1, \ldots, \theta_K) = f(\underline{\theta})$ where $t_j = \sum_{i \in s} w_i \xi_{ji}$.

Berger and Skinner (2005) considered t_j as Hajek (1981) ratio estimator of θ_j. The weight w_i is taken as $\frac{1}{\widehat{N}\pi_i}$ in which $\widehat{N} = \sum_{i \in s} \frac{1}{\pi_i}$.

The mean square error (MSE) of $f(\underline{\theta})$ can be estimated as

$$\widehat{V}(f(\underline{t})) = v(f(t))$$

$$= \sum_{i \in s} \sum_{j \in s} \frac{\pi_{ij} - \pi_i \pi_j}{\pi_{ij}} \epsilon_{(i)} \epsilon_{(j)}$$

denoting $\epsilon_{(i)} = (1 - w_i)\left(f(\underline{t}) - f_{(i)}(\underline{t})\right)$

where $f_{(j)}(t) = f(t_1(j), t_2(j), \ldots, t_K(j))$

and $t_k(j)$ will be calculated as defined in t_j just deleting the jth unit from the sample s.

$$(j = 1, 2, \ldots, n; \quad k = 1, 2, \ldots, K)$$

Here π_{ij} denotes the second order inclusion probability of units i and j $(i \neq j; i = 1, \ldots, N, j = 1, \ldots, N)$.

The term $(1 - w_i)$ in $\widehat{V(f(\underline{t}))}$ is a correction for unequal π_i. The contribution of observations having higher π_i values makes smaller contribution to the variance.

10.3 The Technique of Replicated and Interpenetrating Network of Subsampling (IPNS)

Mahalanobis's (1946) Interpenetrating Network of Subsampling is used in the estimation of variance regardless of the complexity of the form of the estimate, linear or non-linear whatever it may be.

Later (Deming, 1956) named the method as replicated sampling, to get an estimate of variance of an estimator of any parameter which is equivalent to IPNS. The main purpose for the use of the technique was to measure and reduce errors even during the fieldwork and survey procedure. The sample workers were divided into two or more nonoverlapping groups of field workers. Estimate of each part has been calculated. Any awkward divergence among estimates of different groups can create question about the quality of field workers of different groups.

Let K independent samples have been drawn from a finite population with same design of sampling to estimate the population parameter θ. Let $t_1, t_2, \ldots, t_K$ be estimates of θ with $E(t_i) = \theta, i = 1, \ldots, K$. Moreover variance of t_i is same for each sample as the sampling designs are identical in all respect.

So, $V(t_i) = V \quad \forall i = 1, 2, \ldots, K$.

Defining the combined estimator as $\bar{t} = \frac{1}{K} \sum_{i=1}^{K} t_i$, we may write

$$E(\bar{t}) = \theta$$

$$V(f) = \frac{1}{K^2} V \left(\sum_{i=1}^{K} t_i \right)$$

$$= \frac{1}{K} \cdot V \text{ (as they are iid)}$$

The variance may be unbiasedly estimated by

$$\widehat{V}(\bar{t}) = v(\bar{t}) = \frac{1}{K(K-1)} \sum_{i=1}^{K} (t_i - \bar{t})^2,$$

no matter how complicated is the estimator and the sampling scheme.

If $K = 2, \bar{t} = \frac{t_1 + t_2}{2}$.

$$E(\bar{t}) = \theta$$

and $$V(\bar{t}) = \frac{V}{2}.$$

Then $$\widehat{V}(\bar{t}) = v(\bar{t}) = \frac{(t_1 - t_2)^2}{4}.$$

For $K = 2$, it is called half-sampling method.

Statistical independence between samples is vital point here. It may be ensured by the entire replacement of each sample before drawing the next. This technique evolved into the independent form in the first three rounds of the Indian National Sample Survey during October 1950 to November 1951. This is popular because of its simplicity in applications. It is called sample re-use technique. Koop (1967) discussed review of IPNS method. Wolter (1965), Roy and Singh (1973), Chaudhuri and Adhikary (1987) demonstrated different aspects of this method.

10.4 Balanced Repeated Replication

Balanced repeated replication (BRR) is a popular technique for variance estimation in surveys. Under a general stratified, BRR technique can be employed in estimating the finite population mean and the related variance.

Let the finite population of N units be divided into L strata of sizes $N_1, N_2, \ldots, N_L$. The population mean is

$$\bar{Y} = \sum_{h=1}^{L} W_h \bar{Y}_h$$

where $\bar{Y}_h = \frac{1}{N_h} \sum_{i=1}^{N_h} y_{hi}$, $W_h = \frac{N_h}{h}$, $h = 1, 2, \ldots, L$.

y_{hi} is the value of the variable y for hth stratum unit i.

From each stratum, samples of sizes $n_h = 2, h = 1, 2, \ldots, L$ are selected by SRSWOR method.

An unbiased estimator the population mean $\bar{Y}$ is written as

$$\bar{y}_{st} = \sum_{h=1}^{L} W_h \bar{y}_h \text{ where } \bar{y}_h = \frac{y_{h_1} + y_{h_2}}{2}, h = 1, 2, \ldots, L$$

Ignoring the finite population correction (f.p.c) of $V(\bar{y}_{st})$, the estimate of variance of $\bar{y}_{st}$ is

$$v = \widehat{V}(\bar{y}_{st}) = \frac{1}{4} \sum_{h=1}^{L} W_h^2 (y_{h_1} - y_{h_2})^2$$

$$= \frac{1}{4} \sum_{h=1}^{L} W_h^2 d_h^2 \text{ where } d_h = y_{h_1} - y_{h_2}$$

Let the sample (y_{h_1}, y_{h_2}) be divided into two half-samples on taking

$$\delta_h = \begin{cases} 1, & \text{if } y_{h_1} \text{ is the unit of the first half sample} \\ 0, & \text{if } y_{h_2} \text{ is taken in the first half sample} \end{cases}$$

In 2^L possible ways, the half samples can be formed. Then, two unbiased estimators of Y based on first half sample and second half sample are

$$t_{h_1} = \sum_{h=1}^{L} W_h [\delta_{h_j} y_{h_1} + (1 - \delta_{h_j}) y_{h_2}]$$

$$t_{h_2} = \sum_{h=1}^{L} W_h [(1 - \delta_{h_j}) y_{h_1} + \delta_{h_j} y_{h_2}]$$

for jth such formation ($j = 1, 2, \ldots, 2^L$).

$$\text{Then} \qquad \bar{t}_J = \frac{t_{j_1} + t_{j_2}}{2} = \bar{y}_{st}, j = 1, 2, \ldots, 2^L.$$

The variance of $\bar{t}_J$, ie, $V(\bar{t}_J) = V(\bar{y}_{st})$ may be estimated by

$$v_j = \frac{(t_{j_1} - t_{j_2})^2}{4}, j = 1, 2, \ldots, 2^L$$

$$\frac{(t_{j_1} - t_{j_2})^2}{4} = \frac{1}{4}\left(\sum_h W_h \psi_{h_j} d_h\right)^2$$

where $\psi_{h_j} = 2\delta_{h_j} - 1 = \pm 1$, and $d_h = (y_{h_1} - y_{h_2})$, $j = 1, 2, \ldots, 2^L$.
So, $v_j = \frac{1}{4}\sum_h W_h^2 d_h^2 + \frac{1}{4}\sum\sum_{h \neq h'} W_h W_{h'} d_h d_{h'} \psi_{h_j} \psi_{h_{j'}}$.

Summing over all the v_j's ($j = 1, 2, \ldots, 2^L$), its average is defined as

$$\bar{v} = \frac{1}{2^L}\sum_{j=1}^{2^L} v_j = \frac{1}{4}\sum_{h=1}^{L} W_h^2 d_h^2$$

$$= v \text{ as } \sum_j \psi_{h_j} \psi_{h'_j} = 0, j = 1, 2, \ldots, 2^L.$$

But the calculation becomes troublesome as 2^L becomes very large even it for small strata numbers (L).

So we need to form K half samples I and II such that $\sum' \psi_{h_j} \psi_{h'_j} = 0$ where $\sum'$ denotes the sum over this small subset of half sample formations. To form such half-sample replicates, Hadamard matrices with entries $\pm$ (which are square matrices of orders that are multiples of 4) may be used.

To construct above, K should be multiple of 4 and within the range ($L, L + 3$). Thus for $L = 10$ strata, $K = 12$ replicates are required to ensure $\sum' \psi_{h_j} \psi_{h'_j} = 0$.

10.5   Bootstrap

Efron's (1982) bootstrap technique is a computer intensive method for approximating the sampling distribution of any statistic derived from a random sample. It is a resampling technique on computer to extract as much information as possible from the data on hand. A large number of independent resamples or bootstrap samples are drawn from the original sample with replacement. These samples can be used to estimate for any statistic its standard error. For the resample, the non-linear statistic is calculated. Random independent samples of n units are drawn from a random sample of size n to produce an empirical cumulative distribution.

Thus a replica for the unknown true histogram for the original distribution of a random variable is formed. The empirical distribution function $\widehat{F}$ of the original F is obtained.

This is the "plug-in-principle". The bootstrap method is a direct application of the plug-in-principle.

10.5.1 Naive Bootstrap

Let θ be a population parameter which may or may not be linear function of the observations $Y_1, \ldots, Y_N$ considering the units of the population as $1, 2, \ldots, N$.

A sample $s = (1, 2, \ldots, n)$ is drawn by SRSWR method. $\{y_i\}_{i=1}^n$'s are iid yielding $E(y_i) = \bar{Y}$.

An SRS $\{y_i^*\}_{i=1}^n$ of WR size n is drawn from the observed values $y_1, y_2, \ldots, y_n$. It is denoted as s^*. s^* is called a bootstrap sample.

$$\text{Let } \theta = f(\bar{Y})$$

It may be estimated by $\widehat{\theta} = f(\bar{y})$ where $\bar{y}$ is the original sample based estimate of the population mean $\bar{Y}$.
Based on the units of s^*, we calculate

$$\tilde{\theta}^* = f(\bar{y}^*)$$

where $\bar{y}^* = \frac{1}{n} \sum y_i^*, i \in s^*$.

This form of θ includes ratios, regression, correlation coefficient etc.

Repeating the selection of bootstrap samples independently a large no (B) times, we may calculate $\tilde{\theta}^*(1), \tilde{\theta}^*(2), \ldots, \tilde{\theta}^*(B)$.

A bootstrap estimate for θ is taken as

$$\tilde{\theta}_0^* = \frac{1}{B} \sum_{b=1}^{B} \tilde{\theta}^*(b)$$

The bootstrap variance is

$$v_B = \frac{1}{B-1} \sum_{b=1}^{B} \left(\tilde{\theta}^*(b) - \tilde{\theta}_0^* \right)^2.$$

It may be shown that the empirical distribution of $(\tilde{\theta}^*(b) - \widehat{\theta})(b = 1, 2, \ldots, B)$ for large n and B approximates the distribution of $(\widehat{\theta} - \theta)$. So v_B approximates the variance of $\widehat{\theta}$.

The histogram drawn by $\tilde{\theta}^*(b)(b = 1, 2, \ldots, B)$ is a close approximation of the original distribution of $\hat{\theta}$. This histogram is referred as bootstrap histogram. In practice $B = 500$ or 1000 or even 10,000.

Confidence Interval (CI) with a given confidence coefficient (say $100(1 - \alpha)\%$) may be constructed by Percentile method and Double Bootstrap method. Let $100\alpha/2\%$ of the histogram area be below

$$\theta^*_{\alpha/2}, l \text{ and } \theta^*_{\alpha/2}, U.$$

Then $\left[\theta^*_{\alpha/2}, l \text{ and } \theta^*_{\alpha/2}, U \right]$ is taken as $100(1 - \alpha)\%$ confidence interval for θ. This procedure is denoted as percentile method of confidence interval estimation.

In double bootstrap method, a large number of bootstrap samples are drawn from a particular bootstrap sample. Let the bth bootstrap estimate for the parameter θ be $\tilde{\theta}^*(b), b = 1, 2, \ldots, B$. From the bth bootstrap sample, a further set of B bootstrap samples are drawn to get the MSE estimators of $\tilde{\theta}^*(b)$ as given earlier. Denoting the current MSE estimator as $m^*(b)$, let us calculate

$$\delta^*(b) = \frac{\tilde{\theta}^*(b) - \tilde{\theta}^*_0}{\sqrt{m^*(b)}}.$$

With these $\delta^*(b)$ values, a histogram is drawn to get (δ_L, δ_U), where δ_L and δ_U are numbers below and above of 2.5% values of $\tilde{\theta}^*(b)$.

Then the interval $(\hat{\theta} - \delta_L \sqrt{v_B}, \hat{\theta} + \delta_U \sqrt{v_B})$ is a $100(1 - \alpha)\%$ confidence interval under double bootstrap method. An interested reader may consult (Rao and Wu, 1988).

Writing E_* and V_* as the expectation and variance operators with respect to the above bootstrap sampling, then for a large number of B bootstrap samples we may write

$$E_*(\tilde{\theta}^*_0) \simeq \hat{\theta}$$

and
$$V_*(\tilde{\theta}^*_0) \simeq v(\hat{\theta})$$

where $v(\hat{\theta})$ is the original sample (s) based estimate of variance of the statistic $\hat{\theta}$.

For stratified sampling denote

$$\tilde{\theta}^*_{st} = \sum_{h=1}^{L} W_h \bar{y}^*_h$$

where $K = 1$ and $\bar{y}^*_h$ denotes the bootstrap estimate of the mean of hth stratum $h = 1, 2, \ldots, L$.

Then
$$V_*(\tilde{y}_{st}^*) = \sum_{h=1}^{L} W_h^2 V_*(\tilde{y}_h^*)$$

$$= \sum_{h=1}^{L} W_h^2 \left(\frac{n_h - 1}{n_h} \frac{s_h^2}{n_h} \right)$$

where
$$s_h^2 = \frac{1}{n_h - 1} \sum_{i=1}^{n_h} \left(y_{h_i} - \bar{y}_h \right)^2.$$

It is evident that
$$V_*(\tilde{y}_{st}^*) \neq v(\bar{y}_{st})$$

So, $\tilde{\theta}_{st}^*$ is not a good estimator of θ as the bootstrap based variance is not a consistent estimator of the original estimate $v(\bar{y}_{st})$. Efron (1982) calls it a scaling problem of the Naive bootstrap procedure.

10.5.2 Rao and Wu (1988) Rescaling Bootstrap

Rao and Wu (1988) proposed a rescaling bootstrap technique for general unequal probability sampling design. This procedure eliminates the scaling problem of the Naive bootstrap method.

The parameter of interest θ is a non-linear function of K linear totals or means.

$$\theta = f(\theta_1, \theta_2, \ldots, \theta_j, \ldots, \theta_K).$$

A homogeneous linear estimator (HLE) $t_j = \sum_{i \in s} b_{s_i} y_i$ for θ_j is to be employed with b_{s_i}'s as constants and free of Y's.

The MSE is

$$M(t_j) = \sum \sum d_{ij} y_i y_j \text{ with } d_{ij} = E(b_{s_i} I_{s_i} - 1)(b_{s_j} I_{s_j} - 1)$$

where $I_{s_i} = \begin{cases} 1 & \text{if } i \in s \\ 0 & \text{if } i \notin s \end{cases}$.

From Rao (1979), it is known that $M(t_j) = 0$ if $\frac{y_i}{w_i} = $ constant for every i in U. Then $M(t_j)$ may be written as

$$M(t_j) = -\sum_{i<j} \sum d_{ij} w_i w_j \left(\frac{y_i}{w_i} - \frac{y_j}{w_j} \right)^2$$

The MSE estimator is of the form

$$m(t_j) = -\sum_{i<j}\sum d_{s_{ij}} I_{s_{ij}} w_i w_j \left(\frac{y_i}{w_i} - \frac{y_j}{w_j}\right)^2$$

where $d_{s_{ij}}$'s are constants free of y's and $I_{s_{ij}} = I_{s_i} I_{s_j}$ subject to $E(d_{s_{ij}} I_{s_{ij}}) = d_{ij}$ for every i, j in U.

Then (Rao & Wu, 1988) bootstrap procedure to estimate $\theta = f(\theta_1, \ldots, \theta_K)$ and the related MSE may be narrared as follows.

A sample of size n is chosen with the probability $p(s)$ in which the sample size n is fixed. A bootstrap sample is drawn from the sample s as described below.

Out of $n(n-1)$ ordered pairs of units i, $j (i \neq j)$, a bootstrap sample (s^*) of m pairs of units are chosen with replacement (WR) with probabilities $q_{ij} (= q_{ji})$.

Denoting the sampled pair of s^* as $(i^*, j^*)(i^* \neq j^*)$, we define

$$t_j^* = t_j + \frac{1}{m}\sum_{(i^*,j^*)\in s^*}\sum K_{i^*j^*}\left(\frac{y_{i^*}}{w_{i^*}} - \frac{y_{j^*}}{w_{j^*}}\right)$$

where $K_{i^*j^*}$ to be presently specified.

A large number (B) of bootstrap samples (s_b^*, s) is drawn from $n(n-1)$ pairs of the sampled units.

$$E_*(t_j*) = t_j + \sum_{i\neq j\in s}\sum K_{ij}q_{ij}\left(\frac{y_i}{w_i} - \frac{y_j}{w_j}\right)$$

$$= t_j \text{ as } K_{ij}q_{ij} = K_{ji}q_{ji}$$

and
$$V_*(t_j) = \frac{1}{m}\sum_{i\neq j\in s}\sum K_{ij}^2 q_{ij}\left(\frac{y_i}{w_i} - \frac{y_j}{w_j}\right)^2$$

which equals $v(t_j)$ on choosing $m = n(n-1)$ and $q_{ij} = \frac{1}{m}$ and $K_{ij} = \left[\frac{m}{q_{ij}}\left(-\frac{1}{2}d_{ij}(s)w_i w_j\right)\right]^{1/2}$.

Rao and Wu (1988) suggested that the bootstrap based expectation and the bootstrap based variance can be matched to the initial estimate and its variance estimate in the linear case (i.e., for $K = 1$).

Drawing large number of bootstrap samples as above (say B times), the bootstrap estimator of $\theta = f(\theta_1, \ldots, \theta_K)$ can be given by

$$t^* = f(t_1^*, t_2^*, \ldots, t_K^*).$$

So for every linear totals $(t_1, t_2, \ldots, t_K)$, the bootstrap estimators are to be calculated following as above (t_j^*).

The final bootstrap estimator is

$$\bar{t}^* = \frac{1}{B} \sum_{b=1}^{B} t^*(s_b^*)$$

where $t^*(s_b^*) = f(t_1^*, t_2^*, \ldots, t_K^*)$ based on bth bootstrap sample $b = 1, 2, \ldots, B$. The bootstrap estimate of the MSE of the original estimator $t = f(t_1, \ldots, t_K)$ is

$$\tilde{v}_g = \frac{1}{B-1} \sum_{b=1}^{B} \left(t^*(s_b^*) - \bar{t}^* \right)^2.$$

10.5.3 Modifications of Rao and Wu (1988) Rescaling Bootstrap

Rao and Wu (1988) rescaling bootstrap technique was restricted to the fixed size design as the initial drawn sample. Starting with Horvitz-Thompson estimator (HTE) as the linear estimator, Yates and Grundy estimator (YGE) was used. But if the no of units of the initial samples vary across samples, then (Rao & Wu, 1988) rescaling method needs modifications.

Pal (2009) pointed out these amendments and proposed modifications on these. The initial sample size $r(s)$ may vary across samples. In that case, the Horvitz-Thompson (HT) estimator of the linear total $Y = \sum_{i=1}^{N} y_i$ is given by

$$t_H = \sum_{i \in s} \frac{y_i}{\pi_i} (\pi_i > 0) \text{ with its variance as}$$

$$V(t_H) = \sum_{i=1}^{N} \left(\frac{1 - \pi_i}{\pi_i} \right) y_i^2 + \sum \sum_{i \neq j} (\pi_{ij} - \pi_i \pi_j) \frac{y_i y_j}{\pi_i \pi_j}$$

$V(t_H)$ has an unbiased variance estimator

$$v(t_H) = \sum_{i \in s} \left(\frac{1 - \pi_i}{\pi_i^2} \right) y_i^2 + \sum \sum_{i \neq j \in s} \frac{\pi_{ij} - \pi_i \pi_j}{\pi_{ij}} \frac{y_i y_j}{\pi_i \pi_j}$$

(where $\pi_i > 0$ and $\pi_{ij} > 0$).

But unfortunately $v(t_H)$ may turn out negative. In that case we cannot apply (Rao & Wu, 1988) rescaling bootstrap technique to get bootstrap estimate of the non-linear parameter $\theta = f(\theta_1, \ldots, \theta_K)$ and the related confidence interval estimators of $f(t_1, \ldots, t_K)$.

So, to adopt a modified bootstrap technique, we require non-negative variance estimators for the respective estimators of the totals. In that situation (Chaudhuri's, 1981) method of obtaining non-negative variance estimate can be used.

If $v(t_H) < 0$, we need to try to get am estimator e_H for which $v(e_H)$ is positive.

Let $\gamma(s)$ be the effective size of the sample s. One unit of s is dropped and a subsample s' of effective size $\gamma(s) - 1$ is chosen with the conditional probability

$$p(s'|s) = \left(\frac{1}{\gamma(s)} - 1\right) \frac{\sum_{i \in s} y_i}{\sum_{i \in s'} y_i}.$$

The estmator $e_H = \dfrac{\gamma(s) - 1}{\gamma(s)} \left(\dfrac{\sum_{i \in s} y_i}{\sum_{i \in s'} y_i}\right) t_H$ is defined .

$$\text{So } E(e_H) = E(t_H) = \sum_{i=1}^{N} y_i$$

Variance of e_H becomes

$$V(e_H) = V(t_H) + E(W(s)) \text{ where } W(s) = t_H^2 a(s) > 0 \text{ with}$$

$$a(s) = \frac{\gamma(s) - 1}{[\gamma(s)]^2} \left[\left(\sum_{i \in s} y_i\right) \sum_{s' \in s} \frac{1}{\sum_{i \in s'} y_i} - 1 \right]$$

Then $v(e_H) = v(t_H) + w(s)$ is an unbiased estimator of $V(e_H)$.

This approach is repeated till a positive variance estimate is obtained.

$$\text{Here } v(e_H) = \sum_{i \in s} \frac{y_i^2}{\pi_i^2} [(1 - \pi_i) + a(s)] + \sum_{i \neq j \in s} \frac{y_i y_j}{\pi_i \pi_j} \left[\frac{\pi_{ij} - \pi_i \pi_j}{\pi_{ij}} + a(s) \right]$$

$$= E_1 + E_2 \text{ where } E_1 > 0 \text{ and } E_2 > 0.$$

With this positive variance estimate $v(e_H)$, we describe the modifications on Rao and Wu (1988) rescaling bootstrap technique using e_H and $v(e_H)$.

The bootstrap samples are drawn from s' for which $v(e_H)$'s are positive for all K variables.

Following (Pal, 2009), two bootstrap samples s_1^* and s_2^* are independently drawn from s' so that the bootstrap based expectation and the variance will be equal to the original linear estimate (here e_H) and its variance estimate (here $v(e_H)$) respectively.

The first bootstrap sample s_1^* is drawn by Poisson Sampling Scheme as described by Hajek (1958). The details of Poisson Sampling Scheme is described in Chap. 11 of this book.

Let

$$t_1^* = \sum_{i \in s_1^*} \frac{y_i}{\pi_i} \frac{1}{r_i} - t_H + e_H$$

where r_i is the probability of success of the unit i, $0 < r_i < 1$.

Denoting E_* and V_* as bootstrap based expectation and variance operator we can write

$$E_*(t_1^*) = e_H$$

$$V_*(t_1^*) = \sum_{i \in s_1^*} \left(\frac{1}{r_i} - 1 \right) \frac{y_i^2}{\pi_i^2}.$$

This $V_*(t_1^*)$ has to be equated to E_1 to get r_i as

$$r_i = \frac{1}{1 + (1 - \pi_i) + a(s)}, 0 < r_i < 1$$

The second bootstrap sample s_2^* will also be drawn by Poisson Sampling Scheme with Bernoullian trials for each pair of $\gamma(s)(\gamma(s - 1)$ with success probability $q_{ij}, (0 < q_{ij} < 1)$.

Define,
$$t_2^* = \sum_{\substack{(i,j) \in s_2^* \\ i \neq j}} \sqrt{\frac{y_i y_j}{\pi_i \pi_j}} \frac{1}{q_{ij}} - \sum_{i \neq j \in s} \sqrt{\frac{y_i y_j}{\pi_i \pi_j}}$$

Then
$$E_*(t_2) = 0$$

$$V_*(t_2) = \sum_{(i \neq j) \in s} \left(\frac{1}{q_{ij}} - 1 \right) \frac{y_i y_j}{\pi_i \pi_j}.$$

Then
$$q_{ij} = \frac{1}{2 - \frac{\pi_i \pi_j}{\pi_{ij}} + a(s)} \text{ hopefully within } (0, 1).$$

Then
$$V_*(t_2) = E_2$$

So final bootstrap estimator is

$$e_H^* = t_1^* + t_2^*$$
$$E_*(e_H^*) = E_*(t_1^*) + E_*(t_2^*) = e_H + 0 = e_H$$
and
$$V_*(e_H^*) = V_*(t_1^*) + V_*(t_2^*) = E_1 + E_2 = v(e_H)$$

For the other linear totals, the above procedure has to be applied to get the final bootstrap estimator of $\theta = f(\theta_1, \theta_2, \ldots, \theta_K)$.

The bootstrap estimate of θ is

$$e^*(\cdot) = f\left(e_H^*(1), e_H^*(2), \ldots, e_H^*(K) \right)$$

Usually, a large number (B) of bootstrap samples are drawn and the final bootstrap estimator is

$$e^* = \frac{1}{B} \sum_{b=1}^{B} e^*(b), b = 1, 2, \ldots, B$$

The bootstrap variance is

$$v^*(e^*) = \frac{1}{B-1} \sum_{b=1}^{B} \left(e^*(b) - e^*\right)^2 .$$

10.6   Exercises

(i) Show that r_{xy} lies in $[-1, +1]$.

(ii) Although N is known, explain why you employ r_{xy} to estimate using $\sum_{i \in s} \frac{1}{\pi_i}$ as an unbiased estimator for N.

References

Berger, Y. G., & Skinner, C. J. (2005). A jackknife variance estimator for unequal probability sampling. *JRSS, 67*(1), 79–89.

Chaudhuri, A. (1981). Nonnegative unbiased variance estimators. In: Krewski, D., Platek, R., Rao, J.N.K. (Eds.), Current Topics in Survey Sampling. Academic Press, New York, pp. 317–327.

Chaudhuri, A. (2010). *Essentials of Survey Sampling*. Prentice Hall of India.

Chaudhuri, A. (2018). *Survey sampling*. Florida, USA: CRC Press.

Chaudhuri, A., & Adhikary, A. K. (1987). On certain alternative IPNS schemes. *JISAS, 39,* 121–126.

Deming, W. E. (1956). On simplification of sampling design through replication with equal probabilities and without stages. *JASA, 51,* 24–53.

Efron, B. B. (1982). The jackknife, the bootstrap and other resampling plans, 31BM Monograph. No. 38, Philadelphia.

Hájek, J. (1958). Some contributions to the theory of probability sampling. *Bulletin of the International Statistical Institute, 36,* 127–134.

Horvitz, D. G., & Thompson, D. J. (1952). A generalization of sampling without replacement from a finite universe. *JASA, 47,* 663–689.

Koop, J. C. (1967). Replicated (or interpenetrating) samples of unequal sizes. *AMS, 38,* 1142–1147.

Mahalanobis, P. C. (1946). Recent experiments in statistical sampling in the Indian Statistical Institute. *JRSS A, 109,* 325–378.

Pal, S. (2009). Rao and Wu's re-scaling bootstrap modified to achieve extended coverages. *JSPI, 139,* 3552–3558.

Pal, S., & Shaw, P. (2021). Estimation of finite population distribution function of sensitive variable. *Communications in Statistics-Theory and Methods.*

Quenouille, M. H. (1949). Approximate tests of correlation in time series. *JRSS, 11,* 68–84.

Rao, J. N. K. (1979). On deriving mean square errors and other non-negative unbiased estimators in finite population sampling. *JISA, 17,* 125–136.

Rao, J. N. K., & Wu, C. F. J. (1988). Resampling inference with complex survey data. *JASA, 80,* 620–630.

Roy, A. S., & Singh, M. P. (1973). Interpenetrating sub-samples with and without replacement. *Metrika, 20,* 230–239.

Tukey, J. W. (1958). Bias and confidence in not quite large samples (abstract). *AMS, 29,* 614.

Chapter 11
Permanent Random Numbers, Poisson Sampling and Collocated Sampling

11.1 The Permanent Random Number (PRN) Technique

Under PRN sampling, every unit of the population $U = (1, 2, \ldots, i, \ldots, N)$ is independently assigned a permanent random number between 0 and 1, i.e., [0, 1]. Let X_i denote the random number assigned to the unit $i = (1, 2, \ldots, N)$. Then X_i's are arranged in ascending order, i.e., $X_{(1)} \leq X_{(2)} \leq \cdots \leq X_{(N)}$.

The first n units in the ordered list constitute the sample. It is obviously an SRSWOR sample which may be termed as sequential SRSWOR. The last n units, namely $(N - n + 1), \ldots, (N)$ also give another sequential SRSWOR of size n. When new units appear (i.e., births), the frame has to be updated. The units will be labelled as $1, 2, \ldots, N, N + 1, \ldots, N + M$ when M units newly appear. For PRN sampling, a $U[0, 1]$ random number is assigned independently to each birth.

The units which are persisting in the register are called persistants. Births (new) are assigned new PRN. Deaths (closed-down) are withdrawn from the register together with their random numbers. A new sample has been drawn on each sampling occasion using the PRN sampling method.

The purpose of sample coordination is to take into account the samples of previous surveys when drawing a new sample.

Two types of co-ordinations are there in the literature. In negative coordination, the overlap between samples is tried to be minimized. In positive coordination, the overlap is maximized.

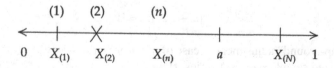

The units are chosen in the first occasion (above) and are labelled on the straight line; a is the starting point. On the second occasion, origin has to be shifted sufficiently

© The Author(s), under exclusive license to Springer Nature Singapore Pte Ltd. 2022   213
A. Chaudhuri and S. Pal, *A Comprehensive Textbook on Sample Surveys*, Indian Statistical
Institute Series, https://doi.org/10.1007/978-981-19-1418-8_11

away from the origin. In order to reduce the overlap between two surveys with sizes n_1 and n_2, we need to choose two constants a_1 and a_2 in $(0, 1)$.

The first sample consists of the units with the h_1 PRN's closest to the right (or left) of a_1. For second sample of size n_2, the PRN's to the right (or left) of a_2 have to be taken.

We may get a negative co-ordination of the samples if we choose a_1, a_2 and the sampling directions properly.

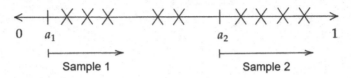

If N is very large, we get a large number of negatively co-ordinated samples.

To get samples with positive co-ordination, the same starting point and direction have to be set.

The estimation procedure will be same as SRSWOR as Ohlsson (1990) described that the sequential SRSWOR is same as ordinary SRSWOR.

Ohlsson also described the PRN techniques for Bernoulli, Poisson and other probability-proportional-to-size sampling procedure.

11.2 PRN Sampling Technique for Poisson Sampling

Poisson sampling was introduced by Hajek (1964, 1971). To draw a sample of size n from the population of N units, the normed size measure is defined as p_i where $\sum_{i=1}^{N} p_i = 1$. Denoting the first order inclusion probability as π_i, we may write $\pi_i = np_i$ and the procedure is referred to as probability proportional to size (pps) sampling.

Brewer et al. (1972) suggested the use of Poisson sampling with permanent random numbers for co-ordination purposes. It can be carried out as follows.

The uniform random number $X_i (X_i \in [0, 1])$ is associated with the unit i. X_i's are mutually independent. The starting point a may be fixed in the interval $[0,1]$.

The unit i will be selected

$$\text{iff } a < x_i \le (a + np_i), i = 1, \ldots, N$$

with a "wrap-around" adjustment in case of $a + np_i > 1$.

The realized sample size $\gamma(s)$ varies. Here

$$E(\gamma(s)) = \sum_{i=1}^{N} \pi_i.$$

Bernoulli Sampling refers to the special case of Poisson sampling where $p_i = \frac{1}{N}, 1, 2, \ldots, N$.

Brewer et al. (1972) suggested the use of Poisson sampling for co-ordination.

Here let X_i be the permanent random number. The positive and negative co-ordination is obtained by choosing appropriate starting point a.

Poisson, Sequential Poisson and Collocated sampling are the only PPS procedures that have been used with PRNs. For Poisson sampling $\pi_i = np_i$, let X_i in the equation

$$a < X_i \leq (a + np_i)(\mod(i))$$

be the PRNs.

The usual (Horvitz and Thompso, 1952) unbiased estimator for the population total

$$Y = \sum_{i=1}^{N} y_i \text{ is } t_H = \sum \frac{y_i}{\pi_i}.$$

Based on Poisson sampling, the variance of t_H is

$$V(t_H) = \sum_{i=1}^{N} \left(\frac{1 - \pi_i}{\pi_i}\right) y_i^2 \text{ as for Poisson sampling } \pi_{ij} = (\pi_i \pi_j, \forall i \neq j)$$

$V(t_H)$ may be estimated as $v(t_H) = \sum_{i \in s} \left(\frac{1-\pi_i}{\pi_i^2}\right) y_i^2$.

The realized sample size $\gamma(s)$ of a Poisson distribution is random. Brewer et al. (1972) recommended that the ratio estimator

$$t_R = \begin{cases} \frac{E(\gamma(s))}{\gamma(s)} \sum_{i \in s} \frac{y_i}{\pi_i} = \frac{n}{\gamma(s)} \sum_{i \in s} \frac{y_i}{\pi_i} & \text{if } \gamma(s) > 0 \\ 0 & \text{if } \gamma(s) = 0 \end{cases}$$

as an alternative to t_H, $\left(n = E(\gamma(s)) = \sum_{i=1}^{N} \pi_i\right)$.

Denoting P_0 as the probability of drawing an empty Poisson sample and the inclusion probability of ith unit as $\pi_{i'} = \pi_i(1 - P_0)$, the second order inclusion probability can be

$$\pi_{ii'} = \pi_i \pi_{i'}(1 - P_0) \text{ where } P_0 = \prod_{i=1}^{N}(1 - \pi_i(1 - P_0)) \text{ and } i \neq i' \in (1, \ldots, N).$$

If Poisson sampling is used, the mean square error of this estimator is

$$V(t_R) \simeq (1 - P_0) \sum_{i=1}^{N} \pi_i (1 - \pi_i) \left(\frac{y_i}{\pi_i} - \frac{Y}{n} \right)^2 + P_0^2.$$

A consistent estimator of $V(t_R)$ is

$$\widehat{V}(t_R) = v(t_R) = (1 + P_0)^{-1}(1 - P_0)\frac{n}{\gamma(s)} \sum_{i=1}^{\gamma(s)}(1 - \pi_i) \left[\frac{y_i}{\pi_i} - \frac{t_R}{n} \right]^2 + P_0 t_R^2.$$

11.3 Sequential Poisson Sampling

In PPS, Sequential Poisson Sampling is used like Sequential SRSWOR.

Let X_i be the random number as stated earlier, $i = 1, 2, \ldots, N$. The normed random number ξ_i is

$$\xi_i = \frac{X_i}{N p_i}, i = 1, 2, \ldots, N.$$

The unit i is included in a Poisson sample with $a = 0$ if and only if $X_i \le np_i$, ie, $\xi_i \le \frac{n}{N}$.

In sequential Poisson sampling, we arrange the units of the population by ξ_i. The first n units are selected from the ordered list. It may be noted that the sequential Poisson sampling reduces to sequential SRSWOR if p_i's are all same. The estimator t_R performs slightly better in sequential Poisson sampling than under Poisson sampling (Ohlsson, 1995).

11.4 Sequentially Deleting Out-of-Scope Units

In survey we may encounter the problem where the list contains many units that are out of scope.

With the sequential technique, we can continue (sequentially) with the random numbers deleting out-of-scope units, until we get n in-scope units.

The probability distribution will be same here as the PRN's are independent. But the number N of in-scope units on the frame is unknown. In that case ratios are being computed with unknown N.

11.5 Collocated Sampling

Collocated sampling resembles Poisson sampling, but the sample size variation in Poisson sampling can be reduced in Collocated sampling. It refers to the collocation of random numbers not the samples.

Here first the population is randomly ordered. Then the random number allocated to the ith unit is $(i = 1, 2, \ldots, N) R_i = .\frac{L_i - \epsilon}{N}$ where ϵ is generated from uniform distribution $[0,1]$.

Then the Poisson sampling scheme is employed just replacing X_i by this $R_i (i = 1, 2, \ldots, N)$. Collocated sampling reduces the probability of an empty sample as R_i's are equally spaced in the interval $[0,1]$.

The estimation formulae for Collocated sampling is same as Poisson sampling.

The similarities between Sequential Poisson and PRN collocated sampling can be evident from their inclusion rule. They are

$$\text{Rank} \left[\frac{X_i}{np_i} \right] \leq n$$

and

$$\frac{\text{Rank}(X_i) - \epsilon}{Np_i} \leq n \text{ respectively.}$$

Both are equivalent to sequential SRSWOR where p_i's are all equal. Chaudhuri (2010) has given details.

11.6 Exercise

Discuss the importance of Poisson scheme of sampling and its special case Bernoulli sampling.

References

Ohlsson, E. (1992). *The system for coordination of samples from the business register at statistics Sweden-A methodological summary, R& D Report 1992:18*. Stockholm: Statistics Sweden.

Ohlsson, E. (1995). Co-ordination of samples using permanent random numbers. In G. Cox, J. A. Binder, B. Nanjamma, A. Christinson, M. J. College, & P. S. Kott (Eds.), *Business survey methods* (pp. 153–170). NY, USA: Wiley.

Chaudhuri, A. (2010). *Essentials of survey sampling*. New Delhi, India: PHI.

Ohlsson, E. (1990). *Sequential poisson sampling from a business register and its application to the swedish consumer price index, R & D Report 1990:6*. Statistics Sweden

Hájek, J. (1964). Asymptotic theory of rejective sampling with varying probabilities from a finite population. *The Annals of Mathematical Statistics, 35*, 1491–1523.

Hájek, J. (1971). Comment on a paper by Basu, D. In V. P. Godambe & D. A. Sprott (Eds.), *Foundations of statistical inference*. Toronto, Canada: Holt, Rinehart & Winston.

Brewer, K. R. W., Early, L. J., & Joyce, S. F. (1972). Selecting several samples from a single population. *AJS, 14*, 231–239.

Horvitz, D. G., & Thompson, D. J. (1952). A generalization of sampling without replacement from a finite universe. *JASA, 47*, 663–689.

Chapter 12
Network and Adaptive Sampling

One of them is 'Network Sampling'. In it two kinds of population units are distinguished—one called 'Selection Units' (SU) with numbers and identities amenable to determination and the second called 'Observation Units' (OU) with numbers unknown and lists unavailable at the start. But the SU's and OU's are interlinked. Probability samples of 'SU's' hence may be chosen by sophisticated techniques and their 'links' with the 'OU's' may be exploited to procure samples of OU's in technically sound manners so as to estimate totals of real variables defined on them.

Using Network samples so selected unbiased estimators of population totals of OU-values and their exact variances and exact unbiased variance-estimators thereof are available as will be narrated in this chapter.

The second technique called 'Adaptive Sampling', on the other hand, is applied when the population though big enough contains relatively rare units bearing characteristics of relevance so that an initial probability sample may not capture enough units bearing features of interest. So, a procedure is needed to enhance the information content by choosing a supplementary sample and the theory of unbiased estimation of population total along with variance expressions and unbiased estimators thereof by proper probability sampling may be applicable. The relevant procedures in this context are reported in details, though in brief.

As explained and illustrated above in Network sampling we need to cover two types of sampling units called Selection units and Observational units for the sake of adequate and enhanced information content in sample survey data. For example, to estimate household expenses on hospital treatment of their inmates owing to specific diseases like thrombosis or gallstone or asthma it is helpful to sample the selection units as hospitals and then sample the households whose inmates have been treated in the selected households for the sake of adequate information rather than from lists of households because many of them may not have inmates at all so suffered and hospital-treated. In usual Sample survey textbooks this topic is hardly covered and students in general are rarely exposed to such specific type of problems and equipped

© The Author(s), under exclusive license to Springer Nature Singapore Pte Ltd. 2022   219
A. Chaudhuri and S. Pal, *A Comprehensive Textbook on Sample Surveys*, Indian Statistical Institute Series, https://doi.org/10.1007/978-981-19-1418-8_12

to deal with such problems when professionally so demanded in more mature lives. Though in the "Network and Adaptive Sampling" book by the senior author the details are covered in the present text we decide to convey the relevant messages to a brief extent.

12.1   Network Sampling: Procedure and Use

Suppose we may intend to estimate household expenses in providing hospital treatment to their inmates suffering from a specific illness, say, cancer or heart disease or brain disease or gall bladder troubles or kidney failures or some such chronic ailment as in-patients in a year in a particular city. Then, by a usual household survey calling at sampled households enough relevant data may be hard to come by because such patients may not exist in most of the sampled households.

It may be a better idea to select a sample of city hospitals and contact the households, the inmates whereof received treatment therefrom for a specific disease of interest in the specific city in the specified time-period, say, a particular year.

Then the treatment centres may be taken as the 'selection units' (SU) and the households linked to them with inmates treated therein for the particular illness needed to be studied, may be identified as the 'observation units' for which the expenses for the inmate-specific treatment may be estimated in respect of the city-households for the year of interest.

Thus, we may take the M treatment centres labelled $j (= 1, \ldots, M)$ in the city as the 'selection units' (SU's) and the households linked to them as the 'observation units' (OU's).

Let a sample of $m (2 \leq m < M)$ SU's be suitably sampled and $A(j)$ be the "set" of all OU's linked to the jth SU and $C_j = |A(j)|$ be the cardinality of $A(j)$, i.e., the "number" of all OU's linked to the jth 'SU' (for $j = 1, \ldots, m$) and m_i be the number of SU's to which the ith OU be linked and let i be an integer extending from 1 through an unknown number N and let y_i be the value of a real variable y of interest like, say, the expenses for hospital-treatment for inmates of the ith household to which the inmates belong, as ascertained from hospitals.

Our interest here is to suitably estimate the total

$$Y = \sum_{i=1}^{N} y_i$$

Let us write

$$w_j = \sum_{i \in A(j)} \frac{y_i}{m_i} \text{ and } W = \sum_{j=1}^{M} w_j$$

Then we get the

Theorem: $\qquad\qquad\qquad\qquad\qquad W = Y.$

Proof:

$$W = \sum_{j=1}^{M} \sum_{i \in A(j)} \left(\frac{y_i}{m_i} \right)$$

$$= \sum_{i=1}^{N} y_i \left(\frac{1}{m_i} \sum_{j|A(j) \ni i} 1 \right) = Y$$

because $\displaystyle\sum_{j|A(j) \ni i} 1$ equals m_i by definition.

Note: So, in order to estimate Y one needs really to estimate W.

Let a sample s of SU's be selected with probability $p(s)$ for an adopted design p.

Let $\qquad\qquad\qquad \pi_j = \sum_{s \ni j} p(s) > 0$

for every $\qquad\qquad j = 1, \ldots, M$

and $\qquad\qquad\qquad \pi_{jj'} = \sum_{s \ni j, j'} p(s) > 0 \,\forall\, j, j' = 1, \ldots, M (j \neq j')$

Then, $\qquad\qquad\qquad t_{HT} = \sum_{j \in s} \frac{w_j}{\pi_j}$

being an unbiased estimator for W may be taken as the Horvitz and Thompson's unbiased estimator for Y as well.

Also, $\qquad V(t_{HT}) = \sum_{j}^{M} \sum_{j'}^{M} (\pi_j \pi_{j'} - \pi_{jj'}) \left(\frac{w_j}{\pi_j} - \frac{w_{j'}}{\pi_{j'}} \right)^2 + \sum \frac{w_j^2}{\pi_j} \beta_j$

writing $\qquad \beta_j = 1 + \frac{1}{\pi_j} \sum_{\substack{j'=1 \\ (j' \neq j)}}^{M} \pi_{jj'} - \sum_{j=1}^{M} \pi_j$

and $\qquad v(t_{HT}) = \sum \sum_{jj' \in s} \frac{(\pi_j \pi_{j'} - \pi_{jj'})}{\pi_{jj'}} \left(\frac{w_j}{\pi_j} - \frac{w_{j'}}{\pi_{j'}} \right)^2 + \sum \frac{w_j^2}{\pi_j^2} \beta_j$

is an unbiased estimator for $V(t_{HT})$.

Any other alternative strategies to estimating W may also be employed equivalently to estimate Y in this context.

12.2 Adaptive Sampling

Suppose we are interested to study the economic conditions of the rural and sub-urban goldsmiths in the state of Gujrat in India. In Gujrat there are many districts and suppose, for example, a big district is chosen and let our first task be to suitably estimate the total number of the goldsmiths only in this district. For this let first a sample of villages and suburban cities/towns be selected. But in the sampled villages and towns enough goldsmiths may not live and work and in many of them there may not exist any at all. But in the entire district there may live and work many indeed. Consequently their combined contribution to the GDP (Gross Domestic Product) may be quite substantial. So, to estimate the total district population of goldsmiths should be adequately estimated because from past surveys and censuses per capita outlay in goldsmithy in a wide region may be well-estimated already and the total number involved in the industry may be in real demand. But an initial sample drawn in a usual survey may not throw-up adequate information content.

In such a situation Adaptive sampling may provide an appropriate relief.

By Adaptive sampling we understand the following.

12.2.1 Adaptive Sampling: Technique and Use

Let $U = (1, \ldots, i, \ldots, N)$ be a finite population of a known number N of identifiable units labelled $i (= 1, \ldots, N)$. Let us be interested in a real-valued variable y with values y_i for i in U. Suppose for many units of U the values y_i may be zero or insignificantly low but the total $Y = \sum_{i=1}^{N} y_i$ may yet be substantial but we 'do' not know beforehand which y_i's may be zero or inadequately valued or which ones may have appreciable values.

Suppose a sample s is selected with a probability $p(s)$ according to a design p but when surveyed only a few y_i's are found with values of noteworthy magnitudes but most of the others insignificantly valued. Conventionally we shall regard some of y_i-values as positive and the rest as zero each. If several sampled y_i's are zeros and only a few are positive, while estimating Y the 'sample data-specific' information content will be treated as meagre and a desire will be to blow it up in a scientific manner. For this what we need first is to define for every unit i of U what we called a 'Neighbourhood', say, denoted by $N(i)$. If the population us territorial, say, a number of villages or towns, we may regard the ith village/town and all the other villages/towns with a common boundary each with it as the neighborhood $N(i)$ of ith unit. The sampled unit i or any other in $N(i)$ may or may not be positive or zero-valued. An adaptive sample $A(s)$ when an initial sample s is drawn and surveyed may be constructed in the following manner.

Suppose a sampled unit i is positive-valued, then examine the y-value of each of the other units in its neighbourhood $N(i)$. If each of the other y's is zero-valued you need to do nothing more. If any of them is positive, then examine each of the latter's neighbour and proceed similarly and stop only when all the neighbours are zero-valued. All the units thus examined starting with i is called the "cluster" of i. Then omitting from the cluster of ith unit the units valued zero for y the set of units in the cluster that remain constitutes the 'Network' of the ith unit, including the ith unit itself as well. Every unit in the cluster of i that is zero-valued is called an Edge unit of ith unit. In the population U every unit is either an edge unit or in a cluster or is a 'single unit cluster'.

By courtesy calling each edge-unit and a 'single unit cluster' also a 'Network', we may easily observe that (1) all the networks are disjoint and (2) all the networks together coincide with the population itself.

An initially drawn sample s from U combined with all the units in the network of each of units of s constitutes an 'Additive' sample $A(s)$ corresponding to s. Let $A(i)$ denote the 'Network' of i and C_i be its 'cardinality' ie, the number of units $A(i)$ contains.

Let $t_i = \frac{1}{C_i} \sum\limits_{j \in A(i)} y_j$ and $T = \sum\limits_{1}^{N} t_i$. Then we have the

Theorem: $T = Y$.

Proof:

$$T = \sum_{i=1}^{N} \left(\frac{1}{C_i} \sum_{j \in A(i)} y_i \right)$$

$$= \sum_{i=1}^{N} y_i \left(\frac{1}{C_i} \sum_{j|A(j) \ni \epsilon_i} 1 \right)$$

$$= \sum_{i=1}^{N} y_i = Y$$

because $\sum\limits_{j|A(j) \ni \epsilon_i} 1$ equals C_i by definition.

Supposing the adopted design p admits

$$\pi_i = \sum_{s \ni i} p(s) > 0 \forall i \in U$$

and

$$\pi_{ii'} = \sum_{s \ni i, i'} p(s) > 0 \forall i, i' \in U,$$

then, $t_{HT} = \sum\limits_{i \in s} \frac{t_i}{\pi_i}$ is an unbiased estimator of $T = \sum\limits_{1}^{N} t_i$ and hence of Y too.

$$V(t_{HT}) = \sum_{i<j}\sum(\pi_i\pi_j - \pi_{ij})\left(\frac{t_i}{\pi_i} - \frac{t_j}{\pi_j}\right)^2$$

$$+ \sum_{i=1}^{N}\frac{t_i^2}{\pi_i}\beta_i$$

writing
$$\beta_i = 1 + \frac{1}{\pi_i}\left(\sum_{j\neq i}\pi_{ij}\right) - \sum_{1}^{N}\pi_i$$

and
$$v(t_{HT}) = \sum_{i<j\in s}\sum\frac{(\pi_i\pi_j - \pi_{ij})}{\pi_{ij}}\left(\frac{t_i}{\pi_i} - \frac{t_j}{\pi_j}\right)^2$$

$$+ \sum_{i\in s}\frac{t_i^2}{\pi_i^2}\beta_i$$

is an unbiased estimator for $V(t_{HT})$.

For $T = Y$ any other procedure for unbiased estimation along with unbiased variance estimator may be obtained if the initial s and derived $A(s)$ may be obtained following alternative sample selection procedures.

Note: For Network sampling as well as for Adaptive sampling the finally effective samples for appropriate estimation of population total may turn out prohibitively enormous. So 'Constraining' efforts are called for in both of them.

12.3   Constrained Network Sampling

Referring to Sect. 12.1, let many of C_j for $j = 1, \ldots, m$ be quite large and $\sum_{j=1}^{m} C_j = C$ be prohibitively large. So, implementing Network sampling in practice may not be practically feasible because of time, money and human and allied resources may be hard to come. So, a somehow constrained Network sampling may be needed to be employed. We may cite below a simple possibility.

Let from $A_j(j = 1, \ldots, m)$ an SRSWOR B_j of size d_j such that $d = \sum_{j=1}^{m} d_j$ with $2 \leq d_j < C_j$ for $j = 1, \ldots, m$, from every A_j for $j = 1, \ldots, m$ be independently chosen across $j = 1, \ldots, m$ on choosing the magnitudes of d_j for $j = 1, \ldots, m$ and d as somewhat manageable.

Let $u_j = \frac{C_j}{d_j}\sum_{i\in B_j}\frac{y_i}{m_i}$ and E_R, V_R denote expectation and variance operators in respect of the above SRSWOR independently across $j = 1, \ldots, m$.

For simplicity, let $a_i = \frac{y_i}{m_i}$. Then

$$V_R(u_j) = \frac{C_j^2 \left(\frac{1}{d_j} - \frac{1}{C_j} \right)}{(C_j - 1)} \sum_{i \in A_j} \left(a_i - \frac{\sum_{i \in A_j} a_i}{C_j} \right)^2$$

and

$$v_R(u_j) = \frac{C_j^2 \left(\frac{1}{d_j} - \frac{1}{C_j} \right)}{(d_j - 1)} \sum_{i \in B_j} \left(a_i - \frac{\sum_{i \in B_j} a_i}{d_j} \right)^2$$

is an unbiased estimator for $V_R(u_j)$. Then, it follows that

$$e = \sum_{j \in s} \frac{u_j}{\pi_j} = t_{HT} \bigg|_{w_j = u_j, j \in s}$$

is an unbiased estimator for Y, because

(i) $E_R(e) = t_{HT}$ and
(ii) $E_p(t_{HT}) = T = Y$ so that

$$E(e) = E_p E_R(e) = E_p(t_{HT}) = T = Y.$$

Then,
$$V(e) = E_p V_R(e) + V_p E_R(e)$$

$$= E_p V_R \left(\sum_{j \in s} \frac{u_j}{\pi_j} \right) + V_p(E_R(e))$$

$$= E_p \left[\sum_{j \in s} \frac{1}{\pi^2} V_R(u_j) \right] + V_p(t_{HT})$$

$$= \sum_{j=1}^{M} \frac{1}{\pi_j} V_R(u_j)$$

$$+ \left[\sum_{j < j'}^{M} \sum^{M} (\pi_j \pi_{j'} - \pi_{jj'}) \left(\frac{w_j}{\pi_j} - \frac{w_{jj'}}{\pi_{jj'}} \right)^2 \right.$$

$$\left. + \sum_{1}^{M} \frac{w_j^2}{\pi_j} \beta_j \right]$$

with
$$\beta_j = 1 + \frac{1}{\pi_j} \sum_{j' \neq j} \pi_{jj'} - \sum_{j=1}^{M} \pi_j$$

and an unbiased estimator for $V(e)$ is

$$v(e) = \sum_{j \in s} \frac{1}{\pi_j} v_R(u_j)$$

$$+ \left[\sum_{j < j' \in s} \left(\frac{\pi_j \pi_{j'} - \pi_{jj'}}{\pi_{jj'}} \right) \left(\frac{u_j}{\pi_j} - \frac{u_{j'}}{\pi_{j'}} \right)^2 \right.$$

$$\left. + \sum_{j \in s} \frac{u_j^2}{\pi_j^2} \beta_j \right]$$

12.4   Constrained Adaptive Sampling

This is a follow-up to our Sect. 12.1 discussed earlier.

If for units i of $U = (1, \ldots, i, \ldots, N)$ the network $A(i)$ turns out too big rendering determination of y_j for j in $A(i)$ very tough, then Constraining Adaptive sampling may be necessary. For this our approach is discussed below.

Let from $A(i)$ an SRSWOR $B(i)$ of sizes d_i for $i \in s$ be selected independently across i in s choosing d_i such that $\sum_{i \in s} d_i = D$, say, such that D is less than $\sum_{i \in s} C_i$, where C_i is the cardinality of $A(i)$ so that surveying the units in $B(i)$'s for i in s may be quite feasible.

Then, let $u_i = \frac{1}{d_i} \sum_{j \in B(i)} y_j, i \in s$. Letting E_R, V_R as operators of expectation, variance in respect of the SRSWOR's above of $B(i)$ from $A(i)$, independently over i in s, we may observe the following:

$$E_R(u_i) = t_i = \frac{1}{C_i} \sum_{j \in A(i)} y_j, i \in U$$

$$V_R(u_i) = \left(\frac{1}{d_i} - \frac{1}{C_i} \right) \frac{1}{C_i - 1} \sum_{j \in A(i)} (y_j - t_i)^2$$

$$= V_i$$

and

$$v_i = \left(\frac{1}{d_i} - \frac{1}{C_i} \right) \frac{1}{d_i - 1} \sum_{j \in B(i)} (y_j - u_i)^2$$

is an unbiased estimator of $V_R(u_i) = V_i$ in the sense that

$$E_R(v_i) = V_R(u_i) = V_i, i \in U.$$

Now, let
$$e = \sum_{i \in s} \frac{u_i}{\pi_i}$$

Then,
$$E_R(e) = t_{HT} = \sum_{i \in s} \frac{t_i}{\pi_i}$$

$$E(e) = E_p E_R(e) = E_p(t_{HT}) = \sum_{i=1}^{N} t_i = T = Y$$

Thus, e above is an unbiased estimator of Y.

Also,
$$V(e) = V_p E_R(e) + E_p V_R(e)$$

$$= V_p(t_{HT}) + E_p \left[\sum_{i \in s} \frac{V_i}{\pi_i^2} \right]$$

$$= \sum_{i < i'} (\pi_i \pi_{i'} - \pi_{ii'}) \left(\frac{t_i}{\pi_i} - \frac{t_{i'}}{\pi_{i'}} \right)^2 + \sum_{1}^{N} \frac{t_i^2}{\pi_i} \beta_i$$

$$+ \sum_{1}^{N} \frac{V_i}{\pi_i}, \beta_i = 1 + \frac{1}{\pi_i} \sum_{i' \neq i} \pi_{ii'} - \sum \pi_i.$$

Then, an unbiased estimator of $V(e)$ is

$$v(e) = \sum_{i < i' \in s} \left(\frac{\pi_i \pi_{i'} - \pi_{ii'}}{\pi_{ii'}} \right) \left(\frac{u_i}{\pi_i} - \frac{u_{i'}}{\pi_{i'}} \right)^2$$

$$+ \sum_{i \in s} \frac{t_i^2}{\pi_i^2} \beta_i + \sum_{i \in s} \frac{v_i}{\pi_i}.$$

A Prediction Approach in Adaptive Sampling

In Adaptive sampling and in Constrained Adaptive Sampling, capturing neighboring rare units turns out difficult because of various hazards. These two approaches can't be implemented in the presence of unobserved units of the network.

Pal and Patra (2021) proposed to try Royall's prediction approach here to model features of uncaptured network units. Brewer-Royall's prediction approach is employed to derive a predictor of the population total with asymptotic design unbiasedness.

Let, the ith sampled unit's network $A(i)$ has a part which is unobserved due to hazardous condition. Denoting $A_s(i)$ as the observed part of the network $A(i)$ and $R_s(i)$ as unobserved part, it can be written as,

$$t_i = \frac{1}{m_i} \sum_{j \in A(i)} y_j = \frac{1}{m_i} \left(\sum_{j \in A_s(i)} y_j + \sum_{j \in R_s(i)} y_j \right).$$

It is obvious that $A_s(i) \cup R_s(i) = A(i) \, \forall \, i \in s.$

Undoubtedly, the second part of the above expression is $\sum\limits_{j \in R_s(i)} y_j$ unknown as these y_j's are unobserved. It follows immediately that t_i is also unknown.

Now the main task is to predict the unknown part $R_s(i)$ of the ith unit's network. A predictor of t_i can be written as,

$$\hat{t}_i = \frac{1}{m_i} \left(\sum_{j \in A_s(i)} y_j + est \left(\sum_{j \in R_s(i)} y_j \right) \right);$$

denoting $est \left(\sum\limits_{j \in R_s(i)} y_j \right)$ the estimate of unknown $\sum\limits_{j \in R_s(i)} y_j$.

A correlation between the units of the same network defined as intraclass correlation may be present and it is denoted by ρ. Considering the model $M : y_j = \beta_i x_j + \varepsilon_j$, where β_i is the unknown constant and ε_j's are random variables having the following features:

$$E_M(\varepsilon_j) = 0, \; V_M(\varepsilon_j) = \sigma_j^2 \text{ (Unknown)}, \; E_M(\varepsilon_j, \varepsilon_k) = \rho \sigma_j \sigma_k$$

where $\begin{array}{l} i=1,2,...,N \\ j,k \in A(i) \; j \neq k \end{array}$,

a predictor of the population total may be derived. E_M, V_M stand for the expectation, variance operators respectively for the model M.

For the simplicity of derivation, Chaudhuri and Stenger (2005) can be followed to approximate the unknown σ_j where $\sigma_j = \sigma_e x_j$, with $\sigma_e \, (> 0)$ as unknown and x_j as known.

Hence, the predictor of t_i is

$$\hat{t}_i = \frac{1}{m_i} \left[\sum_{j \in A_s(i)} y_j + \left(\sum_{j \in A_s(i)} \frac{y_j}{x_j} \left(\frac{1}{\sum_{j \in A_s(i)} 1} \right) \right) \left(\sum_{j \in R_s(i)} x_j \right) \right].$$

The optimum value of the $MSE_i = E_M \left(\hat{t}_i - t_i\right)^2$ is

$$MSE_{io} = \frac{\sigma_e^2}{m_i^2}(1 - \rho) \left\{ \sum_{j \in R_s(i)} x_j^2 + \frac{\left(\sum_{j \in R_s(i)} x_j\right)^2}{\left(\sum_{j \in A_s(i)} 1\right)} \right\}.$$

Under the above model, $e' = \sum_{i \in s} \frac{\hat{t}_i}{\pi_i}$ is an unbiased predictor for $\sum_{i=1}^{N} t_i$ and also for $\sum_{i=1}^{N} y_i$.

Then the unbiased estimator of variance of e' is

$$v(e') = \sum_{i<j \in s} \sum \left(\frac{\pi_i \pi_j - \pi_{ij}}{\pi_{ij}}\right) \left(\frac{\hat{t}_i}{\pi_i} - \frac{\hat{t}_j}{\pi_j}\right)^2$$

$$+ \sum_{i \in s} \frac{v_M\left(\hat{t}_i\right)}{\pi_i^2} - \sum_{i<j \in s} \sum \left(\frac{\pi_i \pi_j - \pi_{ij}}{\pi_{ij}}\right) \left(\frac{v_M\left(\hat{t}_i\right)}{\pi_i^2} - \frac{v_M\left(\hat{t}_j\right)}{\pi_j^2}\right)$$

and $v_M\left(\hat{t}_i\right) = MSE_{io}$, the function of two unknowns σ_e^2 and ρ. These unknowns can be computed by the ANOVA procedure.

12.5 A Brief Review of Literature

In Chapter 6 of the companion book by Chaudhuri (2010) the topics Network Sampling and Adaptive Sampling are discussed in some details. Chaudhuri's (2015) book entitled Network and Adaptive Sampling is a comprehensive text on these topics and so is the one by Salehi and Seber (2013). Thompson (1990, 1992), Thompson and Seber (1996) are other important references and so also are Salehi and Seber's (1997, 2001) works. Chaudhuri (2000) also is a reader-friendly article of note on these two topics. Additional references are Chaudhuri et al. (2004, 2005), Chaudhuri and Pal (2004). One more is Chaudhuri's (2017).

In this text book we do not consider it necessary to discuss here more comprehensibly on these two topics because the coverage on them is principally research-oriented.

12.6   Lessons and Exercises

Lessons

(i) Work out the details on Network and Adaptive Sampling considering sampling and estimation by Rao, Hartley and Cochran's for Network as well as for Adaptive sampling.

(ii) Work out details for Network as well as Adaptive sampling when an initial sample is drawn in two stages.

Exercises

Illustrate two possible cases in practice when (1) Network sampling is applicable, (2) when Adaptive sampling is applicable and (3) Network and Adaptive sampling schemes are applicable in tandem.

References

Pal, S., & Patra, D. (2021). A prediction approach in adaptive sampling. *Metron, 79*, 93–108.

Chaudhuri, A., & Stenger, H. (1992, 2005). *Survey sampling theory and methods* (1st ed.). Marcel Dekker, N.Y.; (2nd ed.). CRC Press, Florida, N.Y. USA

Chaudhuri, A. (2010). *Essentials of survey sampling*. New Delhi, India: PHI.

Chaudhuri, A. (2015). *Network and adaptive sampling*. Florida, USA: CRC Press.

Salehi, M. M., & Seber, G. A. F. (2013). *Adaptive sampling designs*. Springer, Heidelberg: Inference for sparse and clustered population.

Thompson, S. K. (1990). Adaptive cluster sampling. *JASA, 85*, 1050–1059.

Thompson, S. K. (1992). *Sampling*. NY, USA: Wiley.

Thompson, S. K., & Seber, G. A. F. (1996). *Adaptive sampling*. NY, USA: Wiley.

Salehi, M. M., & Seber, G. A. F. (1997). Two-stage adaptive sampling. *Biometrics, 53*, 959–970.

Salehi, M. M., & Seber, G. A. F. (2001). A new proof of Murthy's estimator which applies to sequential sampling. *Australian & New Zealand Journal of Statistics, 43*(3), 901–906.

Chaudhuri, A. (2000). Network and adaptive sampling with unequal probabilities. *CSA Bulletin, 50*, 237–253.

Chaudhuri, A., Bose, M., & Ghosh, J. K. (2004). An application of adaptive sampling to estimate highly localized population segments. *JSPI, 121*, 175–189.

Chaudhuri, A., & Pal, S. (2004). Estimating domain-wise distribution of scarce objects by adaptive sampling ad model-based borrowing of strength. *JISAS, 58*, 136–145.

Chaudhuri, A., Bose, M., & Dihidar, K. (2005). Sample-size restrictive adaptive sampling: An application in estimating localized elements. *JSPI, 134*, 254–267.

Chaudhuri, A. (2017). Unbiased estimation of total rural loans advanced and incurred in an Indian state along with unbiased estimation of variance in estimation. *CSA Bulletin, 69*, 71–75.

Chapter 13
Fixing Size of a Sample in Complex Strategies

13.1 Unequal Probability Sampling and Unbiased Total-Estimator with Well-Known Variance Formula Available

Let t be an unbiased estimator for the total Y with a known form of its variance $V(t)$.
Chebyshev's inequality tells us

$$\text{Prob}[|t - Y| \le \lambda\sqrt{V(t)}] \ge 1 - \frac{1}{\lambda^2}$$

with λ as a constant greater than 1.

Choosing two positive proper fractions f and α let us suppose we need t for Y to be accurate enough that

$$\text{Prob}[|t - Y| \le fY] \ge 1 - \alpha.$$

Combining these two we then need

$$\alpha = \frac{1}{\lambda^2} \text{ and } \lambda\sqrt{V(t)} = fY$$

leading to

$$\alpha = \frac{V(t)}{f^2 Y^2} \tag{13.1.1}$$

Let us postulate a super-population model so as to write

$$y_i = \beta x_i + \epsilon_i, i \in U \tag{13.1.2}$$

© The Author(s), under exclusive license to Springer Nature Singapore Pte Ltd. 2022
A. Chaudhuri and S. Pal, *A Comprehensive Textbook on Sample Surveys*, Indian Statistical Institute Series, https://doi.org/10.1007/978-981-19-1418-8_13

Table 13.1 Calculation of sample-size for PPSWR sampling

N	f	α	σ^2	β	g	n (rounded upto the nearest integer)
100	0.2	0.05	1	5	1.0	22
100	0.2	0.05	1	5	1.5	19
50	0.2	0.05	1	5	1.0	16
50	0.1	0.05	1	10	0.5	12

with β an unknown constant and ϵ_i's as independent random variables with expectation $E_m(\epsilon_i) = 0 \, \forall i$ and variances $V_m(\epsilon_i) = \sigma^2 x_i^g$, $(0 < \sigma < +\infty)$, σ and $g(0 \leq g \leq 2)$ as otherwise unknown.

For simplicity, let us replace (13.1.1) above by

$$\alpha = \frac{E_m V(t)}{f^2 E_m(Y^2)} \tag{13.1.3}$$

Taking for example, x_i's distributed with density $f(x_i) = e^{-z}, 0 < z < +\infty$ and t as Hansen and Hurwitz (1943) estimator denoted as t_{HH}, say, then (13.1.3) gives

$$\alpha = \frac{\sigma^2}{n} \frac{\left[X \sum_1^N x_i^{g-1} - \sum_1^N x_i^g \right]}{f^2 \left(\beta^2 X^2 + \sigma^2 \sum_1^N x_i^g \right)}$$

for PPSWR. Chaudhuri and Dutta (2018) then tabulate the following (Table 13.1).

Here fractions $\frac{n}{N}$ seem OK.

For t as the Horvitz and Thompson (HT) estimator t_{HT} based on an IPPS sample with a fixed size n for each sample, from (13.1.3) one gets

$$\alpha = \frac{\sigma^2}{n} \frac{\left[X \sum_1^N x_i^{g-1} - n \sum_1^N x_i^g \right]}{f^2 \left(\beta^2 X^2 + \sigma^2 \sum_1^N x_i^g \right)}$$

Then, Chaudhuri and Dutta (2018) tabulate (Table 13.2).

Here also $\frac{n}{N}$ values seem OK.

Table 13.2 Calculation of size for IPPS sampling

N	f	α	σ²	β	g	n (rounded upto integer)
100	0.2	0.05	1	10	1.0	7
100	0.1	0.05	1	10	1.0	19
50	0.1	0.05	1	10	1.5	11
50	0.1	0.05	1	10	1.0	13

13.2 Fixing Sample-Size When an Unbiased Variance Estimator is Used

Chaudhuri and Dutta (2019) give the following rules when fixing size of a sample when a sample variance formula is unavailable but an unbiased variance estimator is at hand.

Using Chebyshev's inequality and demanding $|t - Y|$ to be bounded by a fraction of the Y-value with a high probability we noted the rule to use involves the equation

$$100 \frac{\sqrt{V(t)}}{Y} = 100 f \sqrt{\alpha} \text{ with } E(t) = Y$$

If $V(t)$ is not known but an unbiased estimator $v(t)$ for it is at hand (Chaudhuri & Dutta, 2019) prescribes the steps

(1) Take an arbitrary but convenient initial sample-size n for a sampling scheme
(2) Calculate $v(t)$ based on this n
(3) and calculate

$$\frac{100 \sqrt{V(t)}}{t} = \text{CV of } t$$

(4) Obtain a table of values CV of t set against $100 f \sqrt{\alpha} = T$
(5) Take the sample-size as an integer which is a round-up closest to that n for which CV is nearest to the above T.

This is conveniently applicable in the following cases:

(i) Consider SRSWOR combined with (Hartley & Ross's, 1954) unbiased ratio-type estimator for Y as

$$t_{HR} = X\bar{r} + \frac{(N-1)n}{(n-1)}(\bar{y} - \bar{r}\bar{x})$$

with

$$\bar{r} = \frac{1}{n}\sum_{i \in s}\frac{y_i}{x_i}$$

Table 13.3 Table for Hartley-Ross estimator

N	α	f	n	CV	T	Recommended n
67	0.05	0.1	6	15.20	2.236	}6
67	0.05	0.1	11	18.75	2.236	
67	0.05	0.05	17	4.82	1.118	}17
67	0.05	0.05	23	14.78	1.118	
67	0.05	0.05	27	6.07	1.118	

To derive a simple solution for $V(t_{HR})$ is not easy. But an easily derived unbiased variance estimator

$$v(t_{HR}) \text{ is } t_{HR}^2 - \left[\frac{N}{n} \sum_{i \in s} y_i^2 + \frac{N(N-1)}{n(n-1)} \sum_{i \neq j \in s} \sum y_i y_j \right]$$

Applying the above 5 riders (Chaudhuri & Dutta, 2019) give the table below. Here x, y's are derived as in Sect. 13.1 (Table 13.3).

Further for illustration we may consider two more, viz.

(ii) PPSWOR combined with Des Raj estimator and
(iii) PPSWOR combined with symmetrized Des Raj estimator.

An ordered sample $(i_1, i_2, \ldots, i_n) = s$ is chosen with probability

$$p(s) = p_{i_1} \frac{p_{i_2}}{(1 - p_{i_1})} \cdots \frac{p_{i_n}}{(1 - p_{i_1} - \ldots - p_{i_{n-1}})}$$

Des Raj estimator is

$$t_D = \frac{1}{n} \sum_{j=1}^{n} t_j$$

$$t_1 = \frac{y_{i_1}}{p_{i_1}}, \quad t_2 = y_{i_1} + \frac{y_{i_2}}{p_{i_2}}(1 - p_{i_1})$$

$$\vdots \qquad\qquad \vdots \qquad\qquad \vdots$$

$$t_j = y_{i_1} + \ldots + y_{i_{j-1}} + \frac{y_{i_j}}{p_{i_j}}(1 - p_{i_1} - p_{i_2} - \ldots, -p_{i_{j-1}})$$

$$j = 1, \ldots, n$$

An unbiased estimator for $V(t_D)$ is

$$v(t_D) = \frac{1}{2n^2(n-1)} \sum_{j=1}^{n} \sum_{k \neq j}^{n} (t_j - t_k)^2$$

The symmetrized Des Raj estimator is

$$t_{SD}^* = \frac{\sum_{s \to s^*} p(s) t_{D(s)}}{\sum_{s \to s^*} p(s)}$$

and an unbiased estimator of $V(t_{SD}^*)$ is

$$v(t_{SD}^*) = v(t_D) - (t_D - t_{SD}^*)^2$$

Now using $y, x, p_i = \frac{x_i}{X}$ as in Sect. 13.1 and following (Chaudhuri & Dutta's, 2019) 5 rules one may tabulate N, f, α, n, CV and T to recommend appropriate choice of n in practice.

13.3 Fixing Sample-Size When Randomized Response (RR) Survey is to be Designed

Chaudhuri (2020), Chaudhuri and Sen (2020) considered applying Chebyshev's inequality in fixing the sample-size similarly when encountering Randomized Response Survey situations when stigmatizing issues are involved. They illustrated Warner's RRT with SRSWOR and his specimen of quantitative RR scheme. For Warner's qualitative RRT, as usual

$$I_i = 1 \text{ if card type drawn matches } A \text{ or } A^C$$
$$= 0, \text{ otherwise}$$
$$p = \text{Proportion of cards marked } A$$
$$y_i = 1 \text{ if } i \text{ bears } A$$
$$= 0 \text{ if } i \text{ bears } A_C.$$

Then, $I_i^2 = I_i, y_i^2 = y_i$

$$V_R(I_i) = p(1-p), r_i = \frac{I_i - (1-p)}{(2p-1)}$$

$$E_R(r_i) = y_i, V_R(r_i) = \frac{p(1-p)}{(2p-1)^2} = V_i$$

$$\bar{r} = \frac{1}{n} \sum_{i \in s} r_i \text{ has } E_R(\bar{r}) = \bar{y}, E(\bar{r}) = \bar{Y} = \theta, \text{ say.}$$

$$V(\bar{r}) = V_p(\bar{y}) + \frac{p(1-p)}{n(2p-1)^2}$$

$$= \frac{1}{n} \left[\frac{p(1-p)}{(2p-1)^2} + \frac{N}{N-1} \theta(1-\theta) \right] - \frac{\theta(1-\theta)}{N-1}$$

Chaudhuri's RR scheme with quantitative features considers 2 boxes, one with cards marked $a_1, \ldots, a_T$ and the other marked $b_1, \ldots, b_M$ with $\mu_A = \frac{1}{T} \sum_1^T a_j$, $\mu_B = \frac{1}{M} \sum_1^M b_K$,
$\sigma_a^2 = \frac{1}{T-1} \sum_1^T (a_j - \mu_a)^2, \sigma_b^2 = \frac{1}{M-1} \sum_1^M (b_j - \mu_B)^2$.

A respondent draws independently one card from each box to respond

$$I_i = a_j y_i + b_K. \text{ Then}$$
$$E_R(I_i) = \mu_A y_i + \mu_B$$
$$V_R(I_i) = \sigma_a^2 y_i^2 + \sigma_b^2$$

Then,
$$r_i = \frac{I_i - \mu_b}{\mu_a}, \quad E_R(r_i) = y_i$$

$$V_R(r_i) = y_i^2 \left(\frac{\sigma_a^2}{\mu_A^2} \right) + \frac{\sigma_b^2}{\mu_A^2} = V_i, \text{ say}.$$

For SRSWOR,
$$e = \frac{1}{n} \sum_{i \in s} r_i \text{ has } E(e) = \bar{Y} = \theta$$

and
$$V(e) = \left(\frac{1}{n} - \frac{1}{N} \right) \frac{1}{N-1} \sum_1^N (y_i - \bar{Y})^2 + \frac{1}{Nn} \sum_1^N V_i$$

For both the qualitative RRT of Warner and the quantitative case of Chaudhuri's RRT, values of n come out as absurd.

Chaudhuri (2020) further studied these two RRT's to bring out some deep discussions, noting

$$V(e) = E_p V_R(e) + V_p E_R(e)$$
$$= E_p V_R(e) + V_p(t) = I + II.$$

He noted I to dominate II. His conclusion is that n should be fixed concentrating on II only. If thereby I is not appropriately controlled we may not have any relief. This work is a research issue and may not be proper to pursue with it further in this text book on general survey sampling.

13.4   Lessons and Exercises

Writing
$$CV = 100 \frac{S}{\bar{Y}},$$

$$S^2 = \frac{1}{(N-1)} \sum_1^N (y_i - \bar{Y})^2,$$

derive the rules

$$n = \frac{(N-1)(\text{CV})^2}{N\alpha f^2} \text{ for SRSWR}$$

and

$$n = \frac{N}{1 + N\alpha f^2} \frac{(100)^2}{(\text{CV})^2} \text{ for SRSWOR}$$

Postulating Fairfield Smith's model derive the following rule

$$\alpha = \sigma^2 \frac{\sum_n N_i^2 - N}{N(N-1)} \frac{\left[X \sum x_i^{g-1} - \sum x_i^g\right]}{f^2(\beta^2 X^2 + \sigma^2 \sum x_i^g)}$$

which is helpful in settling the sample-size in adopting (Rao et al., 1962) sampling strategy for estimating a population total.

Treating $\frac{t-Y}{\sqrt{V(t)}}$ as a standard normal deviate discuss how you may work out a rule for sample-size determination, especially in case

(i) with replacement and
(ii) without replacement

Hint: See Chaudhuri (2010) and Chaudhuri and Dutta (2018).

References

Hansen, M. H., & Hurwitz, W. N. (1943). On the theory of sampling from finite populations. *The Annals of Mathematical Statistics, 14,* 333–362.

Chaudhuri, A., & Dutta, T. (2018). Determining the size of a sample to take from a finite population. *Statistics and Applications, 16*(1), 37–44. New series

Chaudhuri, A., & Dutta, T. (2019). Pursuing further with an innovative approach the issue to settle the size of a sample to draw from a finite survey population. *Statistics and Applications, 17*(1), 63–72.

Hartley, H. O., & Ross, A. (1954). Unbiased ratio estimators. *Nature, 174,* 270–271.

Chaudhuri, A. (2020). A review on issues of settling the sample-size in surveys: two approaches: Equal and varying probability sampling—crises in sensitive issues. *CSA Bulletin, 72*(1), 7–16.

Chaudhuri, A., & Sen, A. (2020). Fixing the sample-size in direct and randomized response surveys. *JISAS, 74*(3), 201–208.

Rao, J. N. K., Hartley, H. O., & Cochran, W. G. (1962). On a simple procedure of unequal probability sampling without replacement. *JRSS B, 24,* 482–491.

Chaudhuri, A. (2010). *Essentials of survey sampling.* New Delhi, India: PHI.

Chapter 14
Inadequate and Multiple Frame Data and Conditional Inference

14.1 Inadequacy Illustrated of Frames and Way Out Procedures Explained

Network sampling and Adaptive sampling procedures covered in Chap. 12 also are really techniques to get rid of inadequacy in frames. But these are so well-established that a separate chapter is assigned to them and also (Chaudhuri's, 2015) work on them has captured a full book-length thanks to the enterprise by CRC Press, Taylor and Francis.

When strata are fully defined conceptually, say, as disjoint intervals $(a_{h-1}, a_n]$, $h = 1, 2, \ldots, H$ and the units valued in these respective intervals stratified sampling, cluster sampling and multi-stage sampling schemes work if the individuals with values in the respective intervals are specifiable and the lists, if needed, of the units to be selected are prepared. But when the intervals are specified but the units thus valued are not identifiable prior to the sample selection then because of the non-availability of frames stratified sampling is not possible but a way out is Post-stratified sample-survey with a rational approach already discussed earlier.

Chaudhuri (2010a) presented his thoughts on Inadequate Frames in this John Wiley publication, as further elaborated in his book, viz, Chaudhuri (2018). Here we illustrate a different technique to tackle inadequacy in frames described by Chaudhuri (2017).

For several years it was observed that the 'Estimated Amounts' of loans incurred by rural Indians for their productive activities fell very much short of the estimated total loans advanced for the same purpose by various lending agencies to these people, as the respective estimates were determined by Indian National Sample Survey Organization (NSSO) and the Reserve Bank of India (RBI) by dint of their usual sample survey techniques.

The main reason anticipated for this unacceptable discrepancy seems the inadequate frames relied upon by NSSO in its surveys. Chaudhuri (2017) has given a detailed account of efforts to rectify the shortcomings and finally recommended a

© The Author(s), under exclusive license to Springer Nature Singapore Pte Ltd. 2022   239
A. Chaudhuri and S. Pal, *A Comprehensive Textbook on Sample Surveys*, Indian Statistical Institute Series, https://doi.org/10.1007/978-981-19-1418-8_14

procedure described below in some details. He uses Constrained Network sampling scheme as his alternative procedure, vide (Chaudhuri, 2018) essentially employs Constrained Adaptive sampling procedure. His 2017 paper recommends the following in brief. From his 2018 paper we refrain from dealing with here to avoid repetitions.

Chaudhuri (2017) considers his selection units (SU) as the N districts in a given province in a country as the first stage units (fsu) and the M_i lending agencies like banks, co-operatives as the second stage units (ssu) in the ith fsu, $i = 1, 2, \ldots, N$. Out of them $n(2 < n < N)$ fsu's are chosen by Rao et al. (1962) scheme using known census population numbers x_i (and normed size-members $p_i = \frac{x_i}{X}$, $X = \sum_1^N x_i$) are recommended to be drawn followed by $m_i(2 < m_i < M_i)$ ssu's from M_i ssu's are also recommended to be drawn independently by the RHC scheme using the normed size-measures p_{ij} as proportions of known numbers of rural account holders in the concerned lending agencies district-wise.

Let the group-sizes in 1st stage RHC scheme be

$$N_i = \left[\frac{N}{n}\right] + \theta, (= 0, 1) \text{ so that } \sum_n N_i = N,$$

the sum is over the number of groups in RHC sampling of fsu's and

$$M_{ij} = \left[\frac{M_i}{m_i}\right] + p, \, p = 0, 1 \text{ and } \sum_{m_i} M_{ij} = M_i,$$

summing is over the number of m_i groups in RHC sampling of the ssu's.

Let the Account holders of the district-wise lending agencies with unknown numbers and their identities be the Observational Units (OU), labelled $k = 1, 2, \ldots, L$, say with L unknown.

Let A_{ij} be the set of OU's linked to the jth ssu of ith fsu, namely the i, jth SU, with the cardinality $C_{ij} = |A_{ij}|$, ie the number of OU's in the set A_{ij}. Let r_k be the number of SU's to which the kth OU is linked. Let y denote the amount lent out with y_{ij} the same from jth ssu linked to the ith fsu. Then, let $Y_i = \sum_{j=1}^{M_i} y_{ij}$ and

$$Y = \sum_{i=1}^N Y_i = \sum_i \sum_j y_{ij}.$$

Our first problem is to unbiasedly estimate Y. Let z denote the amount borrowed by an OU and z_k be the z-value for kth OU and second problem is to unbiasedly estimate $Z = \sum_{k=1}^L z_k$. In both cases we of course need to unbiasedly estimate the variances of the above two unbiased estimators of Y and Z when suitably derived.

$$\text{Let} \qquad w_{ij} = \sum_{k \in A_{ij}} \left(\frac{z_k}{r_k} \right), j = 1, \ldots, M_i, i = 1, \ldots, N.$$

$$\text{Let} \qquad W = \sum_{i=1}^{N} \sum_{j=1}^{M_i} w_{ij}$$

$$= \sum_{i=1}^{N} \sum_{j=1}^{M_i} \sum_{k \in A_{ij}} \left(\frac{z_k}{r_k} \right)$$

$$= \sum_{k=1}^{L} \left(\frac{z_k}{r_k} \right) \left(\sum_{i=1}^{N} \sum_{j=1}^{M_i} (A_{ij} \ni k) \right)$$

$$= \sum_{k=1}^{L} z_k = Z$$

because the total number of SU's to which the OU's are linked is r_k. So, to estimate Z one may equivalently estimate W.

Letting $A = \frac{\sum_n N_i^2 - N}{N^2 - \sum_n N_i^2}$ and denoting by Q_i the sum of the p_i-values falling in the ith group while applying the RHC scheme of sampling the fsu's,

$$B_i = \frac{\sum_{m_i} M_{ij}^2 - M_i}{M_i^2 - \sum_{m_i} M_{ij}^2}$$

and Q_{ij} the sum of p_{ij}-values falling in the jth group while sampling by RHC scheme the ssu's, m_i in number.

Let $\psi_i = \sum_{m_i} \frac{Q_{ij}}{p_{ij}} y_{ij}$ which is an unbiased estimator of Y_i with an unbiased estimator of its variance as

$$v_i = B_i \sum_{m_i} Q_{ij} \left(\frac{y_{ij}}{p_{ij}} - \psi_i \right)^2$$

An unbiased estimator of Y is then

$$t = \sum_n \frac{Q_i}{p_i} \psi_i = \sum_n \frac{Q_i}{p_i} \sum_{m_i} \frac{Q_{ij}}{p_{ij}} y_{ij}$$

An unbiased estimator of $V(t)$ is

$$v = A \sum_n Q_i \left(\frac{\psi_i}{p_i} - t \right)^2 + \sum_n \frac{Q_i}{p_i} v_i,$$

vide (Chaudhuri, 2010) as discussed in Chap. 3 of this book.

Next applying Network sampling technique it is easy to unbiasedly estimate Z, on identifying A_{ij} for every sampled ij as above and the OU's labelled k linked to the A_{ij}'s as above and then calculate the $w'_{ij}s$. Then, an unbiased estimator for W can be obtained on following exactly the same manner of obtaining t as an unbiased estimator for Y only replacing y_{ij} everywhere by w_{ij} and similarly as v an unbiased estimator of the variance of the unbiased estimator for W as obtained thus as above.

But the crucial problem as should always be encountered if A_{ij}'s be very big and as a consequence surveying all the OU's corresponding to the A_{ij}'s for the sampled ij's may turn out insurmountable.

A way out then is to resort to Constrained Network sampling in the following way.

Let for every sampled ij yielding very big A_{ij}'s with huge C_{ij}'s SRSWOR's B_{ij}'s be chosen independently for respective sampled ij's taking the sizes of B_{ij}'s as d_{ij}'s with $(2 < d_{ij} < C_{ij})$ such that the sum of d_{ij}'s over all the sampled ij's be less than or equal to a modest number d.

Then, let

$$u_{ij} = \frac{C_{ij}}{d_{ij}} \sum_{k \in B_{ij}} \left(\frac{z_k}{r_k} \right)$$

Writing E_R, V_R as operators for expectation, variance in respect of the above SRSWOR, one gets

$$E_R(u_{ij}) = w_{ij}.$$

Then, writing $a_k = \frac{z_k}{r_k}$,

$$V_R(u_{ij}) = \frac{C_{ij}^2 \left(\frac{1}{d_{ij}} - \frac{1}{C_{ij}} \right)}{C_{ij} - 1} \sum_{k \in A_{ij}} \left(a_k - \frac{\sum_{k \in A_{ij}} a_k}{C_{ij}} \right)^2$$

and an unbiased estimator of this is

$$v_R(u_{ij}) = \frac{C_{ij}^2 \left(\frac{1}{d_{ij}} - \frac{1}{C_{ij}} \right)}{d_{ij} - 1} \sum_{k \in B_{ij}} \left(a_k - \frac{\sum_{k \in B_{ij}} a_k}{d_{ij}} \right)^2$$

it now follows that

$$e = \sum_n \frac{Q_i}{p_i} \left(\sum_{m_i} t|_{y_{ij}=u_{ij}} \right)$$

is an unbiased estimator for Z. Hopefully estimated Z should be close to estimated Y because relevant units are linked.

An unbiased estimator for the variance then easily follows as

$$v(e) = v|_{y_{ij}=u_{ij}} + \sum_n \frac{Q_i}{p_i} v_i|_{y_{ij}=u_{ij}} \sum_n \frac{Q_i}{p_i} \sum_{m_i} \frac{Q_{ij}}{p_{ij}} v_R(u_{ij})$$

following variance estimation procedures discussed in Chap. 3 and also in Chaudhuri (2010).

14.2   Multiple Frame Survey Data

Hartley (1962, 1974) illustrated how a population U of N units is listed by one frame A of N_A units ($N_A < N$) combined with another frame B of $N_B(< N)$ units such that $N_A + N_B = N$ but A and B are not disjoint. Rather E_A is a part of A with no units in B, E_B is a part of B with no units in A and AB is composed of the units in A which ate also in B and let N_{EA}, N_{AB}, N_{EB} be the respective cardinalities so that

$$N_{EA} + N_{AB} + N_{EB} = N$$

as E_A, A_B and E_B are mutually disjoint and they together are co-extensive with U.

Let from the frame A an SRSWOR of n_a units be drawn and independently from B another SRSWOR of n_b units be drawn. Further let n_{ab} of n_a units include units also from B and n_{ba} of the n_b units also include units of A. Let, $\bar{y}_a, \bar{y}_{ab}, \bar{y}_{ba}, \bar{y}_b$ be the means of the y-values for a variable y consisting of respectively the n_a units, n_{ab} units, n_{ba} units and n_b units thus sampled.

Choosing a suitable proportion $p(0 < p < 1)$ and writing $q = 1 - p$,

$$t = \frac{N_A}{n_A}(\bar{y}_a + p\bar{y}_{ab}) + \frac{N_B}{n_B}(\bar{y}_b + q\bar{y}_{ba})$$

is an estimator for Y given by Hartley (1962, 1974) and taking a simple cost function and working out the variance of t involving p, q and the variances of y-values of y in E_A, E_B and E_{AB}, he worked out optimum choices of n_A, n_B and $n = n_A + n_B$ and also of p. Saxena, Narain and Srivastava (1984) gave detailed algebra on extending this exercise covering two-stage sampling. Chaudhuri and Stenger (2005) discussed this topic in more details which we choose to omit here.

14.3   Conditional Inference

We are almost through in our plan of producing the present text book. We have covered various approaches in inference-making in the context of survey sampling.

We considered the classical or design-based approach, where we consider a strategy consisting of a sampling design and an estimation procedure. The estimator being a function of the survey data gathered through a survey of the sample-observations based on a sample chosen according to the sampling design adopted. Here the inference is made on calculating the probability-based performance-characteristics reflected through the survey data gathered.

In the model-design based approach and model-assisted approach also conceptual repeated sampling and the probability distribution of the statistics based on the survey-data gathered provide a basis for the inference-making.

In the Predictive approach and Bayesian approach the basis for inference-making is the postulated model and postulated prior and the resulting posterior but the 'survey data' at hand provide the basis for inference-making. No specialization is encouraged about what instead of the current survey data at hand what else would guide the inference by any alternative sample that might have been chosen providing different observations. The sample and the survey-data at hand should of course be of a good quality yielding a quality inference. But the probability distribution of the sample has no role in determining the inferential process. So, here the inference is conditional. The inference is conditioned by the data actually at hand.

But in the contexts of Post-stratified sampling and Small Area estimation inference is often recommended by experts to be Conditional. Let us see how.

In post-stratified sampling a population $U = (1, \ldots, i, \ldots, N)$ of size N, the strata are defined as sets of units with y-values in respective disjoint intervals

$$I_h = (a_{h-1}, a_h], h = 1, \ldots, H$$

$$a_0 = -\infty, a_H < +\infty$$

so that on taking an SRSWOR of n units from U, the units are observed in respect of their y-values and assigned to their respective disjoint intervals called the post strata. Then, a configuration

$$\underline{n} = (n_1, n_2, \ldots, n_h, \ldots, n_H)$$

is determined on observing the numbers n_h found to have come from the respective intervals $I_h, h = 1, 2, \ldots, H$.

Thus, $$E(I_h) = 1 - \frac{\binom{N-N_h}{n}}{\binom{N}{n}}, h = 1, \ldots, H$$

as $$\begin{cases} I_h & = 1 \text{ if } n_h \geq 1 \\ & = 0 \text{ if } n_h = 0 \end{cases}$$

Here N_h is the number of units with values in $I_h, h = 1, \ldots, H$.

Letting $\qquad W_h = \dfrac{N_h}{N}$, and

$$\bar{y}'_h = \frac{1}{n_h} \sum_{i \in I_h} y_i, \text{ if } n_h > 0$$

$$= \bar{Y}_h = \frac{1}{N_h} \sum_{1}^{N_h} y_i, \text{ if } n_h = 0$$

Then, the following t_1, t_2, t_3 are estimators used for $\bar{Y} = \sum\limits_{h=1}^{H} W_h \bar{Y}_h$ which are

$$t_1 = \sum_h W_h \bar{y}'_h$$

$$t_2 = \sum_h W_h \bar{y}_h I_h \Big/ E(I_h)$$

$$\text{and } t_3 = \sum_h W_h \bar{y}_h I_h \Big/ E(I_h) \Big/ \sum W_h \frac{I_h}{E(I_h)}$$

With the configuration

$$\underline{n} = (n_1, \ldots, n_h, \ldots, n_H)$$

at hand (Holt and Smith, 1979) recommend taking the conditional expectations of t_1, t_2, t_3 given the configuration $\underline{n}$ as their respective performance characteristics and also their conditional variances given $\underline{n}$ with calculating the expectations over $\underline{n}$.

The expressions of these conditional expectations and conditional variances with respect to the probability distribution of $\underline{n}$, then these are called unconditional expectations and variances and if these are taken as performance characteristics, then we say we have unconditional inference. If only the conditional expectations and conditional variances are taken as the performance characteristic functions for the estimators used then we say we study only Conditional Inference.

In case of Small Area Estimation also we resort to conditional inference. This is because once a sample is drawn from a population U, observations falling in the respective domains U_d, the domain-sizes N_d and sample-sizes specific to the domains n_d are observed and the statistics employed use the domain-specific units and their values. The expectations, variances of the estimators employed use only the observed quantities n_d, N_d etc. So, here the inference is conditional on the realized values of the domain features.

Rao (1985) is also an important votary of Conditional Inference in Survey Sampling in certain specific circumstances as illustrated above.

14.4   Lessons and Exercises

Work out the exercise in Sect. 14.3 alternatively by dint of Adaptive Sampling instead of Network Sampling Approach.

Solution: See Sect. 7.1 in Chaudhuri (2018), pp. 137–141.

Work out an alternative solution of the problem in Sect. 14.3 employing (Horvitz and Thompson, 1952) estimator in both the stages but choosing the sample of fsu's by the (Hartley & Rao's, 1962) sampling procedure but SRSWOR sampling procedure in choosing the ssu's.

References

Chaudhuri, A. (2015). *Network and adaptive sampling*. Florida, USA: CRC Press.

Chaudhuri, A. (2010a). Estimation with inadequate frames. In B. Benedette, M. Boe, G. Espa, & F. Piersimony (Eds.), *Agricultural survey methods* (pp. 133–138). Chichester: Wiley.

Chaudhuri, A. (2018). *Survey sampling*. Florida, USA: CRC Press.

Chaudhuri, A. (2017). Unbiased estimation of total rural loans advanced and incurred in an Indian state along with unbiased estimation of variance in estimation. *CSA Bulletin, 69*, 71–75.

Rao, J. N. K., Hartley, H. O., & Cochran, W. G. (1962). On a simple procedure of unequal probability sampling without replacement. *JRSS B, 24*, 482–491.

Chaudhuri, A. (2010). *Essentials of survey sampling*. New Delhi, India: PHI.

Hartley, H. O. (1962). *Multiple frame surveys* (pp. 203–206). PSSS SASA.

Hartley, H. O. (1974). Multiple frame methodology and selected applications. *Sankhyā, Ser C, 36*, 99–118.

Saxena, B. C., Narain, P., & Srivastava, A. K. (1984). Multiple frame surveys in two stage sampling. *Sankhyā, 46*, 75–82.

Chaudhuri, A., Bose, M., & Dihidar, K. (2005). Sample-size restrictive adaptive sampling: An application in estimating localized elements. *JSPI, 134*, 254–267.

Holt, D., & Smith, T. M. F. (1979). Post-stratification. *JRSS A, 142*, 33–36.

Rao, J. N. K. (1985). Conditional inference in survey sampling. *Survey Methodology, 11*, 15–31.

Horvitz, D. G., & Thompson, D. J. (1952). A generalization of sampling without replacement from a finite universe. *JASA, 47*, 663–689.

Chapter 15
Study of Analytical Surveys

Goodness of fit test, Test of Independence, Test of homogeneity, Regression and Categorical Data analysis when samples are chosen with unequal selection probabilities.

15.1 Goodness of Fit Test

First we illustrate categorical data. Suppose according to a chosen classificatory character the units of U are each assignable to one of $K + 1$ categories. Thus, let p_i be the probability that a unit of U may be of one of the mutually exclusive categories $i = 1, 2, \ldots, K + 1$. Thus, $0 < p_i < 1, i = 1, \ldots, K + 1$ and $\sum_1^{K+1} p_i = 1$. Let a sample s of size n be drawn from U with probability $p(s)$ according to a design. Then, a frequency distribution of the character may be obtained giving the number of individuals found to have the respective category $i = 1, \ldots, K + 1$. Thus, $\hat{p}_i$ may be derived as consistent estimators for $p_i, i = 1, \ldots, K + 1$.

Suppose, $p_{i0}, i = 1, \ldots, K + 1$ be taken as arbitrary values $p_{i0}, 0 < p_{i0} < 1 \, \forall i = 1, \ldots, K + 1$ such that $\sum_1^{K+1} p_{i0} = 1$. Then, using the $\hat{p}_i$'s and their variances and covariances V_{ij} for $i, j = 1, \ldots, K (i \neq j)$ one may test, using $\hat{p}_i, i = 1, \ldots, K$ and V_{ij}'s or their consistent estimates $\hat{V}_{ij}, i, j = 1, \ldots, K (i \neq j)$ one may test the null hypothesis

$$H_0 : p_i = p_{i0}, i = 1, \ldots, K + 1$$

Let
$$\underline{p} = (p_1, \ldots, p_K),$$
$$\underline{\hat{p}} = (\hat{p}_1, \ldots, \hat{p}_K),$$
$$\underline{p}_0 = (p_{10}, \ldots, p_{K0})$$

For a large n, it is known from standard statistical literature that $\sqrt{n}(\hat{p} - p_0)$ has an asymptotically normal distribution of the k-dimention $N_K(\underline{0}, V)$, where $\underline{0}$ is the k-dimentional vector of 0's and $V = V_{ij}$ is a $k \times k$ matrix of V_{ij}'s, $i, j = 1, 2, \ldots, K (i \neq j)$.

Then, using consistent estimators $\widehat{V}_{ij}$ for V_{ij}, $(i, j = 1, \ldots, K; i \neq j)$ and writing $\hat{V} = (\hat{V}_{ij})_{k \times k}$ it is also known that

$$X_W = n(\hat{p} - p_0)'(\hat{V})^{-1}(\hat{p} - p_0)$$

follows the distribution of a chi-square χ_K^2 with k degrees of freedom (df) if H_0 be true.

So, if the observed value of X_W exceeds $\chi_{K\alpha}^2$ such that

$$\text{Prob}(\chi_K^2 \geq \chi_{K\alpha}^2) = \alpha, 0 < \alpha < 1$$

then X_W provides Wald's test for the above H_0 at a significance level α (usually taken .05), SL.

In general statistical theory observations are generally taken as iid rv's (ie, independently, identically distributed random variables).

Letting Z_i as iid rv's following the normal distribution $N(0, 1)$ with mean zero and unity as standard deviation, then we may observe that $\sum_1^K Z_i^2$ has a chi-square distribution with K as df. So,

$$X_W \sim \chi_K^2 \text{ and } X_W = \sum_1^K Z_i^2.$$

Often it is very hard to find the $\widehat{V}_{ij}$'s unless K is very small. So, an alternative test statistic other than X_W is deemed necessary in practice.

This need is served by Pearson's statistic, namely

$$X_P = n \sum_1^{K+1} (\hat{p}_i - p_{i0})^2 \big/ p_{i0}$$

and the Modified Pearson-statistic

$$X_M = n \sum_1^{K+1} (\hat{p}_i - p_{i0})^2 \big/ \hat{p}_i$$

Both X_P and X_M, for large n are equivalent to each other.

Let us write

$$P = \text{Diag}(\underline{p}) - \underline{p}\,\underline{p}'$$
$$P_0 = \text{Diag}(\underline{p_0}) - \underline{p_0}\underline{p}'_0$$

Then,
$$X_P = n(\hat{\underline{p}} - \underline{p_0})' P_0^{-1}(\hat{\underline{p}} - \underline{p_0})$$

Under H_0, P equals P_0.

If an SRSWR in n draws is taken, then, writing n_i as the observed frequency of units classified into the ith category $(1, \ldots, K)$, then the vector $\underline{n} = (n_1, \ldots, n_i, \ldots, n_K)$ has the multinomial distribution-wise expectation $\underline{p}$ and dispersion matrix P. So, SRSWR in this case is the multinomial sampling.

For general sampling $X_W = \sum_1^K z_i^2 \sim \chi_K^2$ under H_0.

But, for multinomial sampling $X_P \simeq X_M \sim \chi_K^2$ under H_0.

Since X_W is difficult to work out and X_P, X_M are simpler, it is important to check if X_P and X_M, may be further modified to be approximately equated to a X_W under H_0 or examine if they are used as they are, how much deficiency they will suffer from.

Let $D = P_0^{-1} V$ and $\lambda_1, \ldots, \lambda_K$ be its eigen values, namely the roots of the equation

$$|D - \lambda I| = 0, \text{ for a real } \lambda.$$

I is the $K \times K$ identity matrix with units in its K diagonals and 0 everywhere else.

Thus, X_P, under H_0 may be written as $X_P = \sum_1^K \lambda_i z_i^2$.

In case of multinomial sampling $D = I$ and $\lambda_i = 1$ for $1, \ldots, K$ and so $X_P = \sum_1^K z_i^2$ under H_0 and we may reject H_0 if $X_P > \chi_{K\alpha}^2$ achieving α as the SL for the test.

Writing $\underline{c} = (c_1, \ldots, c_K)'$ for c_i as K real numbers, $i = 1, \ldots, K$ $\text{Var}\left(\sum_1^K c_i \hat{p}_i\right) = \underline{c}'V\underline{c}$, for a general sampling design and $\text{Var}\left(\sum_1^K c_i \hat{p}_i\right) = \underline{c}'P\underline{c}$ for the multinomial sampling design .

Recalling
$$\frac{V_{\text{gen}}(\hat{\theta})}{V_{\text{SRSWR}}(\hat{\theta})}$$

for a suitable estimator $\hat{\theta}$ for a parameter θ, as the Design Effect (Deff) given by Kish (1965), the eigen values of $\frac{\underline{c}'V\underline{c}}{\underline{c}'P\underline{c}}$ are called by Rao and Scott (1981) as the "Generalized Design Effects".

Thus, $\lambda_1 = \sup \frac{c'Vc}{c'Pc}$ and (Rao and Scott, 1981) observe that $\frac{X_P}{\lambda_1}$ is approximately equal to χ_K^2 under H_0 and hence $\frac{X_P}{\lambda_1}$ if λ_1 may be roughly guessed provides an approximate test of H_0 with SL close to α.

If an SRSWOR from U is taken of size n, then

$$V = \left(1 - \frac{n}{N}\right) P_0, \, D = \left(1 - \frac{n}{N}\right) I$$

Then, $$P_0^{-1} V = D = \left(1 - \frac{n}{N}\right) I$$

has all the k eigen values $\left(1 - \frac{n}{N}\right)$.

So, for X_P based on SRSWOR $X_P / \left(1 - \frac{n}{N}\right)$ has under H_0 the asymptotic distribution for large n as χ_K^2. So $X_P / \left(1 - \frac{n}{N}\right)$ may be used to test H_0 and if $X_P / \left(1 - \frac{n}{N}\right) \geq \chi_{K\alpha}^2$ then the resulting SL should be close to α and it may be a good test, especially if $\frac{n}{N}$ be quite away from I. What happens for other sampling techniques we do not cover here because of high complexity. But it is better discussed by Chaudhuri and Stenger (2005).

15.2 Test of Independence

Suppose according to two classificatory characteristics A and B the units of a finite population $U = (1, \ldots, i, \ldots, N)$ be classified simultaneously so that a Contingency Table be prepared with the entries in $(r + 1)(C + 1)$ cells as the probabilities $p_{ij} = $ Prob [a unit is i according to A and j according to B].

Then a categorical data analysis is possible to statistically test the null hypothesis H_0 that the two characteristics A and B are independent.

The issue may be formulated as follows

Let $$p_{i0} = \sum_{j=1}^{C+1} p_{ij}, \, p_{0j} = \sum_{j=1}^{r+1} p_{ij}$$
$$h_{ij} = p_{ij} - p_{i0} p_{0j}$$

$\hat{p}_{ij}$ = a consistent estimate of p_{ij} based on a suitable s taken with probability $p(s)$ according to a suitable design p.

Let $p = (p_{11}, \ldots, p_{1C+1}, p_{21}, \ldots, p_{2C+1}, \ldots, p_{r+1C})$
$\underline{h}_I(h_{11}, \ldots, h_{1C}, h_{21}, \ldots, h_{rC})'$
$\underline{p}_r = (\hat{p}_{10}, \ldots, \hat{p}_{r0})', \underline{\hat{p}}_C = (\hat{p}_{01}, \ldots, \hat{p}_{0C}),$

Here $p, \hat{p}$ have $(r + 1)(C + 1) - 1$ elements

and $\underline{\hat{h}} = (\hat{h}_{11}, \ldots, \hat{h}_{21}, \ldots, \hat{h}_{rC})'$ has rC components.

Our object is to test the null hypothesis

$$H_0 : h_{ij} = 0 \, \forall i, j, \ldots i \neq j$$

against the alternative that $h_{ij} \neq 0$ at least for one pair $(i, j), i \neq j$.

Let $H = \frac{\partial h}{\partial p}$, the matrix of partial derivatives of $\underline{h}$ with respect to $\underline{p}$ and $\hat{H}$ be equal to H with every p_{ij} replaced by $\hat{p}_{ij}$ in H.

Let $\frac{V}{n}$ = var-covar matrix of $\hat{\underline{p}}$, $\frac{\hat{V}}{n}$ its estimate, $\frac{1}{n} H'VH$ be the matrix of var-covar of $\hat{\underline{h}}$ and $\frac{1}{n} \hat{H}'\hat{V}\hat{H}$ an estimate thereof.

The Wald statistic for large n to test is then

$$X_W = n\underline{h}'(\hat{H}'\hat{V}\hat{H})^{-1}\underline{h}$$

Under H_0 this X_W is asymptotically distributed as χ_{rc^2} and hence may be used to test this with SL equal to α.

But the Pearson statistic as its alternative is

$$X_P = n\underline{h}'(\text{Kronecker product of} \, \hat{\underline{p}}_r^{-1} \, \text{and} \, \hat{\underline{p}}_C^{-1})\hat{\underline{h}}.$$

For two matrices

$$A_{MXL} = \begin{pmatrix} A_{11} \cdots A_{1L} \\ \vdots \\ A_{M1} \cdots A_{ML} \end{pmatrix}, \quad B_{RS} = \begin{pmatrix} b_{11} \cdots b_{1S} \\ \vdots \\ b_{R1} \cdots b_{RS} \end{pmatrix}$$

their Kronecker Product is

$$A \otimes B = \begin{pmatrix} Ab_{11} \cdots Ab_{1S} \\ \vdots \\ Ab_{R1} \cdots Ab_{RS} \end{pmatrix}$$

Let $T = rc$ and $\delta_1, \delta_2, \ldots, \delta_T$ be the eigen-values of the Kronecker product matrix in X_P.

Then writing z_i's as χ_1^2 distribution for each $i = 1, \ldots, T$, the test statistic X_P for large n under H_0 is distributed as $\sum_1^T \delta_i z_i^2$.

Letting $\delta_1 \geq \delta_2 \geq \ldots \geq \delta_T$, if δ_1 may be guessed approximately then X_P/δ_1 may, for large n, be used to test the H_0 for independence of the two characteristics. But the results X_P/δ_1, however, has the SL less than α and thus it provides a conservative test. But this X_P is appropriate with multinomial sampling and the δ_i's above are interpretable as generalized design effect. In case of goodness of fit test SL for

X_P deviates much from α but this is not so for the test of independence. Simple alternatives to X_P and X_P/δ_1 are hard to come by when general sampling schemes are employed.

15.3   Test of Homogeneity

Suppose the units of two distinct populations be each classified according to a common characteristic A. Let $p_{ji}(0 < p_{ji} < 1, i = 1, \ldots, K+1; j = 1, 2)$ $\sum_{1}^{N} p_{ji} = 1$, for $j = 1, 2$ be the proportions of the classified individuals.

Suppose one sample from each population of sizes n_1 and n_2 be taken by a general sampling design.

Let $\hat{p}_{ji}$ be suitable consistent estimates of p_{ji} based on the respective populations, $j = 1, 2$ for the $(K+1)$ proportions.

Our objective is to use such samples to test the null hypothesis

$$H_0 : \underline{p}_1 = \underline{p}_2 = \underline{p}, \text{ say}$$

Writing
$$\underline{p}_j = (p_{j1}, \ldots, p_{jK})', \underline{p} = (1, \ldots, p_K)'$$

Let
$$P = \mathrm{Diag}(\underline{p}) - \underline{p}\,\underline{p}'$$

$$D_j = P^{-1}V_j, j = 1, 2$$

V_j = variance-covariance matrices of order $K \times K$ of $\hat{p}_{ji}$, $j = 1, 2$ with $\hat{V}_j$ its estimate, $j = 1, 2$

Let
$$\hat{D}_j = P^{-1}\hat{V}_j, j = 1, 2$$

Let
$$\hat{D} = \left(\hat{D}_1/n_1 + \hat{D}_2/n_2\right) \Big/ \left(\frac{1}{n_1} + \frac{1}{n_2}\right)$$

$$\hat{n} = \frac{1}{\frac{1}{n_1} + \frac{1}{n_2}},$$

$$\hat{p}_{0i} = \left(n_1\hat{p}_{1i} + n_2\hat{p}_{2i}\big/(n_1 + n_2)\right)$$

$$\hat{\underline{p}}_0 = (\hat{p}_{01}, \ldots, \hat{p}_{0K})'$$

$$\hat{p}_0 = \mathrm{Diag}(\hat{p}_0) - \hat{p}_0\hat{p}_0'$$

Then the Wald statistic for the above H_0 of homogeneity is

$$X_W = (\underline{p}_1 - \underline{p}_2)' \left(\frac{\hat{V}_1}{n_1} + \frac{\hat{V}_2}{n_2}\right)^{-1} (\hat{p}_1 - \hat{p}_2)$$

the Wald statistic which under H_0 for large n_1 and n_2 follows asymptotically the χ_K^2 distribution.

Thus, $X_W \simeq \sum_1^K z_i^2$, $z_i \sim N(0, 1) \theta_i$, $i = 1, \ldots, K$. The Pearson statistic to test this H_0 of homogeneity is

$$X_P = \bar{n}(\hat{p}_1 - \hat{p}_2)' \hat{P}_0^{-1} (\hat{p}_1 - \hat{p}_2)$$

Let $\lambda_i (i = 1, \ldots, K)$ be the eigen values of $\hat{D}$, the generalized Deff matrix. For large n_1, n_2, under H_0 the statistic X_P is distributed as $\sum_1^K \lambda_i z_i^2$. But to use this X_P it is not easy to work out λ_i's and the SL. Rao and Scott (1981) have studied this problem. Works are scanty for more general sampling than multiple sampling.

15.4 Regression Analysis

Suppose $\underline{Y} = (y_1, \ldots, y_i, \ldots, y_N)'$ represents the N values of a dependent variable y on a finite population $U = (1, \ldots, i, \ldots, N)$. Also let $\underline{X}_N$ be an $N \times r$ matrix giving value of r variables $x_1, \ldots, x_r$ of the N units of U namely x_{ji} for $j = 1, \ldots, r$ and $i = 1, \ldots, N$.

Suppose y and $x_j (j = 1, \ldots, r)$ are related by the Regression relation

$$\underline{r} = \underline{X}_N B^* + \underline{e}N$$

where B^* is a vector of regression coefficients and $\underline{e}N$ the $N \times 1$ vector of error terms with 0 means and constant variances. Then, let

$$B = (\underline{X}'_N \underline{X}_N)^{-1} \underline{X}'_N \underline{Y}$$

be the Least Square estimates of B^* supposing all the N values of y and x_j for $j = 1, \ldots, r$ are available for the entire finite population U.

Now suppose a sample s be selected with a probability $p(s)$ according to a general sampling design p admitting positive inclusion-probabilities π_i and π_{ij} for units i and paired units (i, j) of $U, i \neq j$.

Let $\underline{Y}_s$ be a sub-vector of $\underline{Y}_N$ with only the n units of s of the order $n \times 1$ and $\underline{X}_s$ be the $n \times r$ sub-matrix of $\underline{X}_N$ with x_{ji} values for the units i of U only in s. Let $\underline{W}_N$ be an $N \times N$ matrix of values w_i for i in U and $\underline{W}_s$ and $n \times n$ sub-matrix with element w_i's of $\underline{W}_N$ for i in s. Similarly, let $\underline{\pi}_N$ and $\underline{\pi}_s$ be defined for some real numbers u_i for $i \in U$. Suppose B consists of terms of the form $\sum_1^N u_i w_i$. Then using only sampled quantities let them be replaced by $\sum_{i \in s} \frac{u_i w_i}{\pi_i}$ and terms of the form $\sum_1^N u_i$ be replaced by $\sum_{i \in s} \frac{u_i}{\pi_i}$. Then, using only the sampled values B may be replaced by

$$\hat{\underline{B}}_W = (\underline{X}'_s \underline{W}_s \underline{X}_s)^{-1} (\underline{X}'_s \underline{W}_s \underline{Y}_s)$$

or by
$$\hat{\underline{B}}_{\frac{1}{\pi}} = (\underline{X}'_s \pi_s^{-1} \underline{X}_s)^{-1} (\underline{X}'_s \pi_s^{-1} \underline{Y}_s)$$

Magnitude of $\underline{e}_N$ of course influences $\hat{\underline{B}}_W$, $\hat{\underline{B}}_{\pi^{-1}}$ and the larger it is the worse is the latter.

Further aspects of these may be consulted from Chaudhuri and Stenger (2005).

References

Chaudhuri, A. and Stenger, H. (2005). Survey Sampling: Theory and Methods. 2nd Ed., Chapman and Hall, CRC Press, Taylor & Francis Group, New York.

Kish, L. (1965). *Survey sampling*. NY, USA: Wiley.

Rao, J. N. K., & Scott, A. J. (1981). The analysis of categorical data from complex sample surveys: Chi-squared tests for goodness of fit and independence in two-way tables. *JASA, 76*, 221–230.

An Epilogue

We consider the present treatise as a text book. So, we feel we maintained an easy level in our discussion. We tried to keep it rather reader-friendly avoiding materials tough to grasp. It is after all not a research level composition. But we hope that after assimilating it a curious reader may develop a curiosity to have further readings. Our references are rather comprehensive. But we have not developed materials for a reader to enable him/her to comprehend immediately most of their contents.

I, Arijit Chaudhuri the senior author, who dealt with almost everything in this book except ratio, regression estimators, balanced repeated replication, IPNS and the Chaps. 10 and 11, may disclose the following. In my early formative years I was teaching in an under-graduate college. When I was about to move on to a post-graduate institution, a colleague in the former who happened to be one of my former teachers gave me a piece of unforgettable advice. I should endeavour to leave the students on completing my post-graduate courses on whatever topics taught that any one interested to pursue further studies may immediately take up a Ph.D program to pursue. I took the exhortation to my heart though I doubt if I could succeed in such a mission in my lifelong efforts.

We have presented many formulae for unbiased variance estimators. But how do they fare versus the straight-jacket one namely $m = t^2 - (\hat{Y}^2)$ when $E(t) = \theta$. It is not a research problem for a fresh post-graduate course-leaver studying sampling.

Another one may be how to use survey data gathered from an implemented sampling strategy to estimate the gain or loss in efficiency attainable by a comparable rival sampling strategy. Chaudhuri and Samaddar (2022) in their forthcoming paper have given a somewhat detail account of this.

This book dwelt on procedures of estimating a population total. Hence a procedure follows as a corollary to estimate a population size when it is not known. Hence a mean with an unknown total may be estimated applying ratio estimation procedure. So, one may consider estimating a distribution function for a finite population. Hence one may proceed to estimate a quantile in a finite population. Chaudhuri and Shaw

255

A. Chaudhuri and S. Pal, *A Comprehensive Textbook on Sample Surveys*, Indian Statistical Institute Series, https://doi.org/10.1007/978-981-19-1418-8

(2020) and Pal and Shaw (2021) have given details. We have given the references but not discussed both in the text. This is left for the readers of the text if they are curious about research.

We have pointed out what is a sufficient statistic in a survey population. This leads to the concept of a "complete class" of estimators of a finite population total. Any estimator in a class outside the complete class in it is composed of 'inadmissible estimators' worse than the estimates in the complete class. So, one should be curious if there may exist 'Admissible' estimators. In a class of estimators an estimator is 'Admissible' if there exists no estimator 'better' than it. In our companion book 'Essentials of Survey Sampling' (cf. Chaudhuri (2010)) details about how to find an 'Admissible' estimator among all unbiased estimators for a finite population total are described following (Godambe and Joshi, Godambe and Joshi (1965)) in their paper Admissibility and Bayes estimation in sampling finite population—AMS 36, 1707–1730.

A reader of the present text on assimilating it should be curious enough to peruse the material with interest. We choose to avoid here needless repetition of this material in Chaudhuri (2014) text on pp. 66–68, as the material is rather tough to be covered in this text intended to be generally accessible.

As a reviewer of the present text resents the non-coverage of the topic called "Outliers in Surveys" it behoves us to pay attention to this grumble. The book *"Outliers in Statistical Data"* by Vic Barnett and Tobby Lewis (John Wiley, 1994) is an authentic text on "Outliers". We know from here that in the context of a finite survey population an outlier is a unit with a value either a very small proportion of units valued still less or valued still higher. In our study of a finite population we have to live with every individual who is a giant in terms of his/her gigantic resourcefulness or with negligible number of social followers so long as they exist as matters of fact. We cannot afford to dispense with them in our society. A multimillionaire is as important for us as is a destitute is valued to us in our studies of social importance. In course of carrying on field survey of course it is important to decide if an abnormally low or high value has been encountered inducing doubt about the genuineness of the sample finding. Such outliers must be examined and decisions appropriately have to be taken regarding their genuineness in being a correct representative of the population. So, we have to sift the values correctly for a corresponding sampling unit to be retained or discarded as a truly representative member of the population.

References

Chaudhuri, A., & Samaddar, S. (2022). Estimating gains in efficiency of complex over simple sampling strategies and studying relative model variance of pair-wise rival strategies with simulated illustrations. Statistics in Transition, New Series, Accepted on 31 March, 2021 for publication in 2022.

Chaudhuri, A., & Shaw, P. (2020). A finite population quantile estimation by unequal probability sampling. *Communications in Statistics-Theory and Methods, 49*(22), 5419–5426.

Pal, S., & Shaw, P. (2021). Estimation of finite population distribution function of sensitive variable. *Communications in Statistics-Theory and Methods*.

Chaudhuri, A. (2010). *Essentials of survey sampling*. New Delhi, India: PHI.

Godambe, V. P., & Joshi, V. M. (1965). Admissibility and Bayes estimation in sampling finite populations I. *AMS, 36*, 1707–1722.

Chaudhuri, A. (2014). *Modern survey sampling*. Florida, USA: CRC Press.

Printed in the United States
by Baker & Taylor Publisher Services.

Printed in the United States
by Baker & Taylor Publisher Services